Nationalism

Nationalism provides a comprehensive exploration of nationalist identity, ideology, and practice which centers the geographic underpinnings of the phenomenon. It unpacks the fundamental principles and the many variations of this global phenomenon, as it examines nationalism through a spatial lens.

Nationalism is *the* dominant political force in the modern world and no other global ideology is so strongly tied to concepts like territory, homeland, frontiers, and boundaries. The authors delve into how nationalism is fundamentally related to territory and place, why mapping is critical to the nationalist endeavors, the role of performance and personification, ethnonationalism, multinationalism, nationalist movements, and how nationalism is evidenced and experienced in cities and towns throughout the world. These provide a solid summary of what makes nationalism so compelling, so uniting, and so dangerous. *Nationalism* provides a fresh and compelling perspective on a complicated and often controversial subject.

Written in an accessible and attractive style, the book will be especially useful for classes in Geography, Global Studies, International Relations, Political Science, Sociology, History, and Anthropology. It provides information and conceptual insights to scholars interested in a concise and sophisticated synthesis of contemporary nationalism. For casual readers interested in the phenomenon of nationalism, this book provides clear explanations and compelling examples.

David H. Kaplan is a Professor of Geography at Kent State University. He has written some 70 peer-reviewed articles and chapters, and has also published *Segregation in Cities, Nested Identities, Boundaries and Place, Human Geography, Urban Geography, Landscapes of the Ethnic Economy, Perthes World Atlas*, the four-volume *Nations and Nationalism: A Global Historical Overview, Navigating Ethnicity*, and *Scaling Identities*. Dr. Kaplan's research interests include nationalism, borderlands, ethnic and racial segregation, urban and regional development, housing finance, and sustainable transportation. Dr. Kaplan has supervised 55 graduate students and teaches courses on many different aspects of human geography. He is also the past President of the American Association of Geographers. He edits the *Geographical Review*, the flagship journal of the American Geographical Society, as well as *National Identities*.

Kathryn Hannum is an Assistant Professor of Geography at Michigan Tech University. Her research interests include national identity, migration and diaspora, sociolinguistics, memorial landscapes, and geographies of sport. She teaches a myriad of courses and leads study-away programs that connect lived experiences to complex Geographic concepts, such as nationalism, migration, and tourism. Dr. Hannum has written several book chapters, as well as articles published in *The Professional Geographer, GeoHumanities*, and *Journal of Cultural Geography*, among others.

Key Ideas in Geography

Series editors: Noel Castree, *University of Wollongong* and **Audrey Kobayashi,** *Queen's University*

The Key Ideas in Geography series will provide strong, original and accessible texts on important spatial concepts for academics and students working in the fields of geography, sociology and anthropology, as well as the interdisciplinary fields of urban and rural studies, development and cultural studies. Each text will locate a key idea within its traditions of thought, provide grounds for understanding its various usages and meanings, and offer critical discussion of the contribution of relevant authors and thinkers.

Mobility, 2nd Edition
Peter Adey

City, 2nd Edition
Phil Hubbard

Resilience
Kevin Grove

Postcolonialism
Tariq Jazeel

Non-representational Theory
Paul Simpson

Climate Change
Mike Hulme

Space
Peter Merriman

Home, 2nd Edition
Alison Blunt and Robyn Dowling

Nationalism
David H. Kaplan and Kathryn Hannum

For more information about this series, please visit: www.routledge.com/series/KIG

Nationalism

David H. Kaplan and Kathryn Hannum

Routledge
Taylor & Francis Group
LONDON AND NEW YORK

Designed cover image: © David H. Kaplan

First published 2024
by Routledge
4 Park Square, Milton Park, Abingdon, Oxon OX14 4RN

and by Routledge
605 Third Avenue, New York, NY 10158

Routledge is an imprint of the Taylor & Francis Group, an informa business

British Library Cataloguing-in-Publication Data
A catalogue record for this book is available from the British Library

Library of Congress Cataloging-in-Publication Data
Names: Kaplan, David H., 1960- author. | Hannum, Kathryn, author.
Title: Nationalism / David H. Kaplan and Kathryn Hannum.
Description: First edition. | New York : Routledge, [2023] |
Series: Key ideas in geography | Includes bibliographical references and index.
Identifiers: LCCN 2023027494 (print) | LCCN 2023027495 (ebook) |
ISBN 9781138350601 (hbk) | ISBN 9781138350618 (pbk) | ISBN 9780429435782 (ebk)
Subjects: LCSH: Nationalism. | Historical geography.
Classification: LCC JC311 .K285 2023 (print) | LCC JC311 (ebook) |
DDC 320.54–dc23/eng/20230729
LC record available at https://lccn.loc.gov/2023027494
LC ebook record available at https://lccn.loc.gov/2023027495

ISBN: 978-1-138-35060-1 (hbk)
ISBN: 978-1-138-35061-8 (pbk)
ISBN: 978-0-429-43578-2 (ebk)

DOI: 10.4324/9780429435782

Typeset in Sabon
by Taylor & Francis Books

Contents

Figures

Boxes

Preface

Like so many book projects, this book came about because we perceived a true gap in the literature on nationalism. Nationalism is such a dominant force in the modern world that it is easy to forget that it is a distinct ideology. It is also easy to forget that nationalism is a fundamentally geographical idea. No other global ideology focuses quite as heavily on geographic concepts like territory, homeland, frontiers, and boundaries. No other global phenomenon is so closely linked with the art and science of mapping. Both of us had tried to unpack the fundamental principles and the multifarious variations of nationalism and, as geographers, we were accustomed to looking at phenomena through a spatial lens. Yet while we located several books on nationalism – written mostly by historians, political scientists, sociologists, even philosophers – there was no single source which examined nationalism geographically. Geographers have offered up some solid research monographs, and led several excellent collections and handbooks, but no singular geographic voice was available. The absence of a comprehensive book which focuses on the key link between geography and nationalism was a glaring omission in the field.

That is why we decided to write a book which would uncover what we saw as the key elements of nationalism from a geographical perspective. Nationalism lacks the coherence of many ideologies; it varies greatly depending on the specific contexts in which it's encountered. Sometimes nationalism surrounds civic engagement, other times it arises out of fear of others. A single national group can span borders, include a multitude of cultures, languages, and citizenships, and have contested history internal to the group. Amid this ideological chaos, there are clear drivers and outcomes which are worthy of discussion. As geographers, we are accustomed to speaking across disciplinary and national boundaries. We attempt to illustrate the thematic richness of nationalism with case studies which span the world, keeping in mind geography's two most fundamental concepts: space and time. Within this, we felt we could utilize geographical notions of territory, landscape, scale, and mapping to help make sense of this variety.

This book contains a few elements seldom found in books about nationalism. First, the book's holistic, coherent approach moves fluently across space, from one era to another, and between places and scales, showing their peculiarity and interconnectedness. Second, the book has a unique, high-quality emphasis on visualization and concrete illustrative examples. Third, the book is both accessible and attractive in style and based on rigorous research. Fourth, the book engages with everyday life, which adds to its value in the classroom and in the eyes of professional readers outside academia.

For students in upper division undergraduate and graduate levels, this book offers conceptual clarity and innovative pedagogy. This will be especially useful for classes in

Geography, Global Studies, International Relations, Political Science, Sociology, History, and Anthropology. For scholars interested in a concise and sophisticated synthesis of contemporary nationalism, this book will facilitate interdisciplinary communication. We tried to bring in a wide range of sources, and the accompanying bibliography will serve as a benchmark for readers interested in pursuing the topic further. Audiences will find high-quality visualization of the examined phenomenon appealing, due to strong traditions of cartography and information design in these realms. For journalists seeking an understanding of this significant phenomenon, this book will combine rigorous scholarship with understandable and accessible prose and illustrations. For casual readers interested in the phenomenon of nationalism, this book will provide a connection between academic concepts and everyday landscapes, events, and symbols.

In terms of the writing and production of the book, Dave found a willing publisher in Routledge and began working with the team in 2017. Kathryn was brought in as a co-author in 2021 and helped bring the book to completion. In addition to the two principal authors, we had help in developing the textboxes from two of our colleagues. Michaela Gawrys was primarily responsible for the textboxes on Unification Nationalism, Symbolism in Flags, Nationalism in the Military, and Ukrainian Breakaway Regions. Michaela also provided some photos and line art. Tiera Moore was primarily responsible for the textboxes on Soviet Identity and Nationalism in Biafra. Both Michaela and Tiera contributed to the textboxes on Nunavut Identity and Indigenous Nationalism in Colombia. We thank them both for their tremendous work.

In addition, we would like to thank the people at Routledge for their help in moving the project forward: Andrew Mould, Egle Zigaite, and Claire Maloney. Pauliina Raento helped to frame some of the initial ideas in the book's structure and Noel Castree offered a considerable number of helpful suggestions. Daniel Lazzadro McPhearson provided valued cartographic consultation, and Stephanie Burkey quickly produced final maps which are attractive and effective. Veronica Jurgena and Mark Rhodes both offered moral support during long writing stretches.

Chapter 1

Variations of Nationalism

In 1863, during the throes of the American Civil War, a story appeared in *The Atlantic Monthly* (Hale 1863). Titled "The Man Without a Country," it told of an Army lieutenant who was tried for treason alongside Aaron Burr and, in a moment of passion, exclaimed that he never wanted to hear of the United States again. The lieutenant, Philip Nolan, was granted his wish and was forbidden to set foot on American soil, and it was decreed that nobody should mention the United States or provide him with news. Instead, he was kept on navy boats at sea, physically separated from his national home. In the story, Nolan grows increasingly desperate to hear news of the United States and despairs of his sentence. He tells the story's narrator, "remember, boy, that behind all these men you have to do with, behind officers, and government, and people even, there is the Country Herself, your Country, and that you belong to Her as you belong to your own mother." Nolan finally receives news of the United States and all of its changes while he is dying, excepting "this infernal Rebellion."

It was no accident that such an expression of unbridled nationalism was written in the midst of the Civil War. Beyond the carnage, the very concept of American national identity was being challenged. A patriotic and nationalistic message, one that included both Union and Confederate states, was an important bridge to a much more cohesive identity than had existed before. The "country" Nolan came to long for was not a single state or a group of people, but instead was expressed as the entire American territory. His dying wish granted to hear of the news was in hearing about just how much the land he had left in 1807 had grown.

The creation of an "American" nation and national identity is a perfect example of the modern essence of nationalism. Being a relatively new country, the political dominion the United States had over the territory it claimed was only as strong as the cultural continuity contained within. Lacking in Western understandings of traditional culture, mythical past, common ethnic heritage or even a congruent landscape that those in power could understand, intellectual observers at the time sought to ground the country in national tradition (Bronner 2002). So, a vast nation building campaign ensued, deploying new folk figures like Davy Crocket, and mythologies like manifest destiny in a successful attempt at uniting people under one flag and in a common political mission.

The national history of the United States is one of hundreds of stories told around the world, as peoples became members of a nation and spaces became national territory. It is partly because there are so many individual stories to tell that it becomes difficult to arrive at a comprehensive theory of nationalism. As John Breuilly (2013) says, nationalism can claim a universal validity, and in that sense, it is similar to liberalism, conservatism, or socialism, yet

DOI: 10.4324/9780429435782-1

its universalism is predicated on the unique character of each nation. So, the study of nationalism lacks many of the sweeping transhistorical referents that unify other ideologies.

There have been dozens of wonderful books on nationalism and several comprehensive readers, handbooks, and encyclopedias. All of these publications must grapple with the fact that nationalism is difficult to pin down beyond its core tenets, but they also open up the myriad ways in which nationalism is experienced. Nationalism is a political doctrine based on the primacy of cultural affinity. Its emergence results in changes in the economic and social attributes of a people, and its success has resulted in the most profound organization of society of the modern era.

Nationalism and Geographic Themes

One area that has been less explored has been how nationalism and national identity relate to geography. That is not to minimize the dozens of geographers writing significant pieces on nationalism, not to mention the hundreds of others who have tackled concepts critical to nationalism. It is simply that there is not yet a single text that considers nationalism from a geographical perspective. For this, we need to consider concepts such as territory, scale, landscape, networks, borders, and cartography.

Nationalism is an essentially *territorial* concept. The goal of a national movement is to acquire self-determination over a piece of land, and by doing so, invest it with territorial properties. This works on both an emotional and a practical level. Territory is how nationalists describe their nation, using terms such as homeland, motherland, fatherland, even "land" itself. Territoriality is the manner by which members of a nation meld themselves to a political state with sovereignty over a clearly bounded space. As Paasi (1996) points out, individuals are socialized around territorial principles. Even members of a nation dispersed from their ancestral land, such as members of a diaspora, are anchored to the prospect of some sort of territorial consolidation in the future.

Territory is able to control and shape space and what resides within a space. This is especially germane in maintaining a nation. Cultural variations, both large and small, will exist within any space. A territorial conception of this space can diminish many of these differences, while augmenting the differences between the nation within this territory and whatever is outside. State control of a national territory comes with control of the educational apparatus, bureaucracy, commerce, and all sorts of things which strengthen national identity. State control of territory is also an impediment to rival national groups who must work within an organizational milieu which is stacked against them. However, claims to some aspect of territory, like a district or a province, can help a sub-state nation promote its own national vision.

Yet territoriality alone can be a conceptual trap in our understanding, as articulated most clearly by Agnew (1994). People do not always operate within two-dimensional units of sovereign space but are instead imbued within *networks*. These networks can be constituted by any number of things that affect national identities: flows of capital, labor, communications, goods, international relationships, non-state powers, immigration, and emigration. Networks as a cascading set of hierarchical relationships are what preceded the emergence of territorial states. Networks, whether hierarchical or unranked, continue to frame national actions. Consider the significance of transnationalism which ties together people from a locality in one country to a destination community in another country.

Nationalism shows up time and again on the *landscape*, and each nation can point to distinct landscape elements. This reflects how natural and human features figure into the way nations conceive of themselves. The national land itself – the coast, the rivers, various islands, forests, and other features – provides important notes. These elements are often featured in patriotic songs or as parts of the flags, coins, stamps, or other national badges. The "islandness" of some nations might confer feelings of protection and exclusivity. Major rivers flowing across the national territory can unite a nation as well, allowing for interaction and communication.

The landscape of the built environment is as important as the natural landscapes. Every nation includes big ticket monuments, buildings, and other structures that consciously define it. Some are of long duration, some were built as a testament to national greatness, some became imbued with a national character later on. Some more commonplace, even ubiquitous landscape elements, communicate nationalism constantly. Flags flying, pictures of leaders (especially common in authoritarian countries), and the musical soundscape of anthems and patriotic songs are constant reminders. Unintentionally national elements may also carry weight, from the language on signage, to the house styles, types of restaurants and shops, the way fields are plowed and crops grown, and just about any element that may communicate national identity (Edensor 2020).

Landscape elements also differ within a country, demonstrating internal regionalism and sometimes nationalism. Milder regional difference may be displayed by features unique to that place, reminiscent of Vidal de la Blache's consideration of *pays*. Travel within a country will reveal many little variations. When a country is divided into two or more distinct cultures, major national difference will be clearly evident in all aspects of the discursive landscape, languages, churches, architecture, and even flags. Often such differences are highlighted as they have much to do with the particular goals and aspirations of national groups.

Scale is a key attribute of nationalism and national identity. For one thing, we can see national identity as part of a host of identities, each of which is situated at a specific scale – larger than a "regional" scale but smaller than a "civilizational" or "global" scale. There are several ways to consider it, whether as steps on a ladder or as a series of ever-expanding circles (Herod 2010). We also must acknowledge that national identities are nested such that a smaller identity is contained within a larger identity and that identity may in turn be contained within a yet larger identity. While a strong nationalist agenda would discourage loyalties at larger or smaller scales and demands singular loyalty to a nation-state at a single geographic scale, identity and loyalty operate in more complicated ways.

Several wrinkles apply to considerations of national scale. First, the scale of any entity is not immanent; rather its construction depends on circumstances. This nation's scale does not remain static. It expands and diminishes over time and diverges among its people. An expansive nation may welcome all newcomers who join in its mission. A more constrained nation limits itself to a smaller population who meet required ethnic and racial criteria. Second, the construction process depends on many actors. The state government has enormous influence and often sees itself as enforcing the nation's integrity, but other actors work either in concert, in opposition, or at cross-purposes to state goals. Third, the national scale must always be considered in light of other identities at both higher and lower scales. The interplay between identities at various scales can create tension and potential conflict. Also, within a given state, the identity of one group of people may situate at a separate scale from the identity of another group, creating what Herb (2004) calls a "double vision."

For geographers, scale is fundamental to spatial inquiry, yet the idea of scale becomes more salient to everyday people and lived experiences through application and imposition of *borders*. Borders come in multiple forms, from physical to imagined, superimposed to antecedent, relic to planned. Political borders regulate the flow of people, goods, services, and ideas across space, making them incredibly important geopolitical demarcations. Borders are a spatial manifestation of power, traditionally used as a device of regulation and surveillance. Because of this, borders can be spaces of tension or war, as when two groups come into contact with one another, there may be conflict. But borders can also be spaces of bridging, where national cultures mix and multiply.

Despite globalization and an increasing liberalization of movement across borders, borders remain very important symbols with very real effects. Those in power are more likely to engage in "border work" than others (Jones & Johnson 2016), and those with little power or capital have much more difficulty crossing borders. In a way, political borders represent the ultimate manifestation of national membership, legally and spatially delineating who is on the inside and who is outside of the nation in the most literal sense.

The bordering of nations occurs in more ways than just the political. Borders emerge between national groups as boundaries defined by national traits. These boundaries can manifest as language use, clothing styles, food consumption patterns, and even the color that one chooses to paint their home. These boundaries are iterated and re-iterated through performances that communicate understanding of national tradition within the group and build symbolic boundaries around national groups. These social boundaries are important for sub-state or transborder groups, groups who must assert their unique status outside of the homogenizing forces of a territorially bounded state.

Maps are the Geographer's ultimate tool, as they communicate the at times abstract ideas elaborated on in this book, in a visual meant to be understood by anyone. Maps use borders, territory, and scale as their cartographic building blocks, and as such convey the importance of national landscapes and national groups. Maps by necessity generalize the real world, and the people in it. It is due to this generalization that maps can work so well as a political tool. Maps can work to obscure some national identities while favoring others. They can be made to strike fear or insecurity into viewers. They can show data that rallies a national cry of war or that quells a population to peace. Maps are powerful and can be powerfully manipulative.

To exist on the map is to be a recognized sovereign territory, the ultimate goal for many national groups. Those interlocking, color filled shapes that fill out continental landmasses are an ultimate symbol of national identity in and of themselves. Shapes of national territories become so recognized and commonplace that they are successfully deployed on everything from stickers to physical borders, to clothing, literally filling the national landscape with reminders of the nation's existence on the map, unique among the shapes that surround it. When displayed like this, as a solid territorial entity, whether floating in space or among geopolitical neighbors, a national territory on a map seems firm, ever-present, and eternal.

Approach and Layout of this Book

Fundamentally nationalism is a geographical ideology and national identity has a strong geographical base. However, there are so many aspects of nationalism and so many individual stories that it is difficult to put all of it into one neat container. Scholars who develop a more comprehensive perspective of nationalism have categorized nationalism into different types (see Kohn 1944) or grasped the various fundamental factors leading to a national identity (see Smith 1986, Anderson 1991, Gellner 2003), or simply proceeded by examining individual nations, as is the manner taken by several encyclopedic works (see Herb and Kaplan 2008 or Breuilly 2013).

All of these approaches have value but what we are trying to do here is something a little different. First, we take an approach that looks at how nationalism and national identity operate in several different spheres of activity. So, we consider the role of the various components that make up the nationalist ideology or which demonstrate how nationalism is manifested. Second, we are attentive to scale, examining national identity as it exists at scales other than that of the state. This assumes a view of the local scale, the regional scale, as well as the supranational scale. Third, we make sure to explore nationalism from the standpoint of geography. Thus, the attention to territory, landscapes, borders, networks, and depictions of these elements, all of which have strong geographical implications.

Like any book that covers a global phenomenon, the burden is on us to provide a number of good examples from around the world. As an ideology, nationalism began in Europe and so this means that many illustrations come from this continent. But it has been useful to see how other countries and nations outside of Europe have adopted nationalism, contended with their own challenges, and added their own elements to how national identity is constructed.

In regard to how this book is organized, Chapter 2 examines the evolution of the nationalist ideology over time and how nationalism is constructed. We look at the role of key philosophers and how they laid out the tenets of modern nationalism. Then we examine the material conditions underlying nationalism, which swiftly displaced other modes of state organization. As nationalism spread from Europe, it was adopted in several types of societies often emerging from a colonial past. Under this perspective, nationalism could be viewed as both a legacy of European influence and as an anti-colonialist strategy. Finally, we examine the significance of ethnonationalism in the framing of national identity.

In Chapter 3, we look specifically at the importance of territory to national identity and the practice of nationalism. The use of certain territorial strategies is discussed in developing borders and in territorializing the national vision. The discursive value of the landscape to the nation demonstrates how nationalism is continuously expressed, even in banal ways, and how sometimes the elements of the landscape are contested. The geographical relationship between state and nation can lead to several problems, as the two are not usually in complete alignment. This follows with the decision by states to handle their minority populations, often in repressive and violent ways. Finally, we look at the importance of transnationalism and diasporas as networks cutting across country borders.

In Chapter 4, we delve into the ways in which nationalism is propagated using symbolism, ephemera, and propaganda. Using illustrative examples, we discuss how the boundaries of the nation are represented in spatial illustration, through atlas and wall maps, logo maps, and the cartographic choices that are used to elicit emotion in the reader and solidify the idea of a unified territory. Discussion then moves to other symbols that represent and fill out the boundaries of the nation, though objects that circulate and are passed around within the everyday landscapes that citizens traverse. We show how banal objects such as stamps, money, clothing, and graffiti can be used as pervasive tools of the nation. This chapter also explores popular geopolitics through cartoons, comics, music, and literature.

Next, we turn to performance, agency, and embodiment of the nation in Chapter 5. Here we look at some everyday reproductions of the nation through forms of doing: ritualistic consumption, participation, collective action, memorialization, and celebrations. We take a look at everyday performances from driving, going to restaurants, and dancing, to more punctuated celebrations that reproduce the nation, like parades, holidays, and global events that center national difference, like the Olympic Games. In this chapter we also look at how national values and beliefs have been represented through allegorical characters or personifications of the nation itself, and how these characters are consumed, celebrated, and reproduced.

Chapter 6 deals with the issue of how states handle multinationalism within their borders. Since a true nation-state is rare, diversity is a feature of countries around the world. A rough schema is introduced that classifies different types of multinational states, and then we look into how national and state identities at different scales can coexist or not. The structure of the state, whether unitary or federal, and the degree to which minority nations are given political rights (in theory and practice) is covered. Finally, we look at the phenomenon of supranationalism, which creates an additional layer of identification at a scale higher than that of the state.

Political attempts to harness minority national movements are often unsuccessful and Chapter 7 discusses the variety of separatist national movements and what are some of the factors in their success and failure. When countries are diverse, the interaction of culture and region plays a significant role. We delineate aspects of separatist movements and what motivates their formation, highlighting the importance of relative costs, the geopolitical situation, and changing identities. We also look at those special instances where a national movement crosses state borders, bringing in a whole other level of complexity.

Nationalism is most attentive to scale, and in Chapter 8 we discuss how nationalism operates at a local scale. While often fighting over larger questions of identity and territory, national movements and divisions occur in local places. Many cities throughout the world contain major fissures between groups; some of these have exploded into vicious conflict. Beyond this, cities are an important component of a nation's view of itself. Capital cities assume outsized importance, and the transformation and creation of capital cities are important national projects. The urban landscape is also tremendously symbolic, and here national visions can be seen in how the urban landscape is shaped.

Chapter 9, our conclusion, wraps up what we see as the major geographical takeaways of nationalism in the world today. The themes presented in this chapter are reexamined and we also look towards where nationalism might go in the future. Nationalism has been described as Janus-faced, hydra-headed, and a phenomenon which is complex because it relies on each unique case (Breuilly 2013). We hope that this book will provide more clarity to this elusive concept.

Chapter 2

Growth of an Ideology

One of history's greatest pivots occurred as two revolutions rocked the late 18th century. Unlike previous wars, the American Revolution (1775–1783) was not a battle between rulers but a battle for self-determination of the ruled. It refuted the notion that a monarch had a legitimate claim to land that had been previously conquered. Earlier wars had been for territory and resources, as royals pressed their subjects to fight battles for greater power. But this was different. The creation of the United States from separate colonial provinces was the first undertaking to forge a new nation, one that barely held together on its own at first but through history acquired more of a strong "national" identity (McCardell 2008). And though the founders of the new United States represented the elite of agriculture, trade, and industry, they put down a marker that a nation of people – however so defined – had the right to self-governance.

About a decade later, the French Revolution of 1789 also was fought on behalf of the people, this time against everything the monarchy stood for. Whereas the American Revolution was a successful effort to free a colony from an empire's grasp, the French Revolution occurred within the territory of an existing state. It was instigated by those with little social and economic status, against not only the monarchy but against the entire hierarchical structure of French society. To French revolutionaries, the nation as represented by the people should have ownership of the state apparatus (Taras 2008). The Revolution sought to overturn existing society completely, introducing equality, ending feudalism, promoting a secular society, and other ideas that were considered quite radical at the time. The aftermath of the French Revolution was politically unstable, leading to the rise of Napoleon Bonaparte and a new French Empire, but many of its tenets have persisted to the present day.

Both the American and French revolutions insistently introduced nationalism to the world. What followed in the 19th and 20th centuries were a series of movements, consolidations, and revolutions that could all rightfully be called "nationalist" in sharp distinction to what had come before. In this chapter, we look at some of the philosophical and material foundations of the nationalist ideology and examine how it diffused across the world.

Nationalism as Ideology

Today nationalism affects how we organize the world and how we frame our experiences. It is so ubiquitous as to mostly be invisible. But nationalism is also an ideology – a body of ideas used to legitimize or challenge authority (Eagleton 1991). These ideas are

DOI: 10.4324/9780429435782-2

represented in a system of signs, of the meanings within language that promote a particular way of acting in the world.

Some ideologies may be quite visible inasmuch as they represent a body of beliefs different from our experience. Within a capitalist system, Marxism or socialism would be seen as quite ideological because these offer a competing worldview. When an ideology is commonplace, when it is successful, we do not consider it especially ideological. Since most people implicitly endorse the ideology, they are less aware of its existence.

Nationalism is one such commonplace ideology. It forms a pervasive aspect of modern life, subscribed to by nearly all peoples in the world. While it contains many variations, these all point to the same ideas: the descriptive idea that the world is divided into nations and the normative idea that these nations deserve to determine their own fate (Breuilly 1985). It would be hard to find a world leader who did not implicitly subscribe to this philosophy, and it would be hard to find a citizenry not swayed by nationalist ideals. Countries as distinct as North Korea and the United States – contrasting in every other conceivable way – are animated by nationalism.

While the ideology is so commonplace, the expression of nationalism is often derided. The bad view of nationalism derives from its most virulent forms. Examples of heavy-handed nationalist movements are legion – from the fascist dictatorships of the 1920s–1940s to the Balkan Wars of the 1990s to the right-wing populism of today's Hungary. Nationalism in these instances equates with exclusion, intolerance, oppression, genocide. But nationalism can also be celebrated as a force for good. Nationalism animates the sentiment undergirding much of the modern welfare state – the desire to take care of one's own. It reveals itself in pictures of immigrants taking the Pledge of Allegiance for the first time. Whatever its reputation, nationalism underlies our modern world system, even our social psychology. It manifests itself left and right, exclusive and inclusive, outward-looking and inward-seeking. And it is always there, buttressing our most popular sentiments and justifying governmental action. The constant invocations to "the [insert name of nation here] people" demonstrates this time and again.

Yet if we were to go back 500 years, nationalism would have little ideological sway. Nationalism relies on national identity, and national identity relies on the sense of belonging to a common people whom you cannot personally know. There was no such sense of national commonality and so no real ideal of national self-determination. Instead, people would be ruled by powers with little cultural affinity. In fact, they might not even speak their language. In pre-national societies, literacy was rare and those able to read and write did so in languages separate from what commoners spoke. We might even describe more of a two-track system within pre-national Europe (Burke 2013). At one level were the elites who belonged to a larger community than any nation. This was often defined by religious communities as well as an extended aristocracy. In regard to religion, the sacred texts would be read by priests and interpreted by them. In terms of politics, the language of the courts could be dramatically different from any national language. At the other level were the vast peasant majority, whose community was their region made up of culturally similar villages and small-scale society. The languages they spoke were pre-literate and peculiar to just a small patch of land.

The pre-nationalist world exhibited many cultural distinctions based on language, religion, folkways, and other factors. It is just that these differences did not map into political states. At a larger scale, they were based on religion and lines of nobility, while at a smaller scale they were localized, fluid, and not tied to any political identity (Tilly 2002). Within Europe, Catholicism forged the linkage between an individual and the wider realm. New Protestant movements challenged Catholicism's sway and tended towards less universality but retained a connection of community through religion (Marsh 2008). In pre-national history, most allegiances were vertical, towards a ruler, a priest, and soldier. Of course, there would be feelings of loyalty to family and fellow villagers, but this was not elevated into a political loyalty. The Medieval conception of "fealty" was reserved for one's lord and master.

Another feature of the pre-nationalist world would be in the relation towards a people and territory. As Herb (2018) points out, before the 15th century, authority was based on personal relations. Allegiances between people and laws were determined by the ruler for his or her subjects. With the exception of the early mercantile separation of the city from the feudal countryside, territory played little role in this and only came into being as the modern state system came to formation.

As an ideology, nationalism is embedded in geography. Nationalism has the effect of strengthening divisions between various peoples, transforming what might have been a continuum of cultural expression into sharp breaks between socially and spatially defined cultural communities. Because it is tied to the development of a strong state system, nationalism encourages the construction of strong borders and the idea that a territory should form the basis of political and cultural power. Geography is also implicated in the textures of the landscape itself. Nationalism valorizes particular places and populates both city and countryside with symbols of national identity. Nationalism may also use the landscape itself as a form of symbolism – drawing strength from certain "typical" elements. National anthems will often exalt the mountains, plains, lakes, seacoast, forests, or any other elements that bind a nation together. Coinage and stamps may represent some places or landscape features prominently.

Philosophical Underpinnings of Nationalism

Nationalism is both a movement and an ideology justified by several historical thinkers. It has no single towering figure, as is true with other ideologies, but rather several people who have developed tenets of nationalism specific to their own experience and time. It is also true that the same philosophers could change their points of view when confronted with new challenges (Dahbour and Ishay 1995). As a result, nationalism is not a singular doctrine but one with several facets. This reflects what nationalism represents (Breuilly 2013). Nationalist ideology speaks to the unique character of each nation. So, it is difficult to devise an overarching idea that is transnational except one that justifies the existence of nationalist consciousness and the imperative that this consciousness be somehow actualized.

It is open to debate just when people began to frame a nationalist worldview. While tracts promoting these ideals flourished in the 19th century, there were a few discussions related to nationalism in the 18th century. The French philosopher Jean-Jacques Rousseau has been particularly emphasized, inasmuch as he wrote about the necessity of cultural homogeneity, achieved through national education and sports, and the importance of a national mindset and a national language (Qvortrup 2003). Beyond the cultural attributes, Rousseau was also a proponent of the feeling of horizontal loyalty, one that people would share a bond with one another if they lived in the same country and shared cultural characteristics. In his *Dissertation on Political Economy* (1758), he wrote: "how can [people] love their [country], if their country be nothing more to them than to strangers" (in Bowden 2003). Duty to the people as a whole, as opposed to some higher power, was a critical aspect of Rousseau's philosophy and an essential part of the nationalist idea.

Proto-nationalist thought fit under the broad penumbra of nationalism, but later philosophers diverged in their views of nationalism. One point of division was whether nationalism can be a truly universalistic and cosmopolitan ideology or whether it was parochial and particularistic. The issue of whether nationalism is more universal or parochial lies at the heart of how nationalist ideology presents itself to others. Rousseau's vision was unclear, in part universalistic or partly parochial depending on the source (Miscevik 2008). Yet other philosophers clearly staked a claim on either side of this divide.

One of the earliest philosophers, Johann Gottfried Herder, developed the thesis that nations were defined by natural characteristics, in particular language. Herder called for cultural self-determination, but he also propounded the idea that national cultures could peacefully co-exist and that one was not superior to the other. In fact, he railed against the imposition of the French language over the European courts of the late 18th century and called on the German language communities to resist (Benner 2013). And while Herder likely coined the term "Nationalismus" he did not see the ultimate expression of national distinction as culminating in the occupance of a political state; he saw the cultural and political as separate spheres (Berlin 1976). And in a 1765 address, Herder stated that the growth of a universalizing Christianity made nationalism less salient (Berlin 1976).

Herder's views were complicated and do not map easily onto later conceptions of nationalism. Like many later nationalists, Herder viewed cultural differences organically, as an inalienable aspect of human identity (Eley and Suny 1996). His work acknowledged and valorized differences as they existed and called for these differences to be tolerated (Bowden 2003). At the same time, Herder was not in favor of the kind of cultural-political project which defined most nationalisms, and he was also pessimistic that strong cultural differences could exist side by side within a single political entity (Benner 2013).

A more parochial view of nationalism came into favor not long after Herder. The best-known proponent of this view was Johann Fichte. Like Herder, Fichte viewed nations as natural, organic entities. In his most famous work, *Addresses to the German Nation* (1807/2008), Fichte spoke of a "fatherland" handed down over generations, within which it was the duty of every German to sacrifice for this entity: "In order to save his nation he must be ready even to die that it may live." In two major ways Fichte deviates from Herder. First lies in the notion that the cultural goals of a nation must be achieved through a state. So even though state and nation are separable, the interests of the nation must take priority over everything else, which then follows that the attainment of national goals are fulfilled through political sovereignty (Miscevic 2008). Second is the idea of the moral superiority of the German people who were sent to lead humanity (Randall 1965). What makes this address remarkable is that it occurred at a time when Germany did not exist as a unified state, but as a collection of duchies, counties, and cities (Applegate 2008); unification would not occur until more than 60 years later, in 1871.

Box 2.1 Unification Nationalism

Figure 2B.1 Unification nationalism
Source: Frédéric Sorrieu, Public domain, via Wikimedia Commons

As states became more defined as territorial units, there was a concomitant push for a more cohesive polity. The unification of peoples who lived within these units, far less considered before, became far more urgent (Kochin 2018). Creating a unified polity provided the states with a number of advantages, including a mobilized citizenry, less internal conflict, a more cohesive economic unit, and greater internal legitimacy. States in turn used different tactics to further promote citizen identification (Brown 1998).

Many of the tools needed for a unified country were eagerly promoted by state governments. A common language, a shared history, distrust of outsiders: these and other elements created a strong state-centric identity by fostering a unified national identity (McLane 2008; Saatci 2002). Symbols such as flags, monuments, maps, and national heroes provide sources of common identity that were previously nonexistent (Cox 2021). A designated national language demands conformity by quickly creating barriers between those who can and cannot participate (Anderson 1991). A unified nationalism encourages the development of an urgent "us vs. them" mentality. State boundaries also promote unification, as people are placed within an international framework that legally marks off difference. These tactics establish affinity with the ingroup and dislike for the outgroup, so there is a lot of pride in identifying with the "correct side" (Allport 1979).

The positive aspects of a state nationalism harness more cohesive state power and internal legitimacy. However, attempts to unify all members of a state under a single national umbrella are rarely fully successful. Some people either are not allowed membership or decide to pursue their own identity. Many such disenfranchised groups develop an alternate nationalism that threatens the state (Gellner 2003). With perceived threats from within and without, the state may feel its very existence compromised and go to extremes so as to maintain power (Blaney and Inayatullah 2000). The symbols of state nationalism are amplified as a result.

These two notions of nationalism inform debates related to larger supranational entities such as the European Union or to parochial sentiments expressed by some current world leaders. These also have sometimes co-existed within the same worldview. For instance, the great Italian politician, Giuseppe Mazzini, argued for the significance of national differences and labored to fuse an Italian state out of the shared nationality of the Italian peoples. At the same time, he saw national differences as transcended in pan-European harmony (Dahbour and Ishay 1995).

A second debate concerns whether national membership is involuntary or if it can be considered as more a shifting synthesis that may reflect a voluntary belonging. Views like those promulgated by Fichte strongly argued for national identity as something that could not be denied. Nations were natural and organic. Romantic images of nations stemmed from something almost metaphysical, beyond human decision making. Later 19th-century writer Heinrich von Treitschke argued that just as the nation must make up the state, this bond must be based on a "natural bond of blood-relationship" (Guibernau 2013, 11).

However, many proponents of nationalism argue for a more voluntary arrangement. Rousseau himself saw the nation as a grouping of people subscribing to the political authority that they themselves create (Benner 2013). In the 19th century both the Englishman John Stuart Mill and Frenchman Ernest Renan argued for national identity as something which was based on the shared desires of the people. Mill saw nations as a voluntary grouping by feeling. In his view, multiple and varied factors could be responsible for a national sentiment, including descent, language, religion, geography, or a shared history which he considered the most likely. None of these factors was "indispensable" as he put it (Mill 1998/1859). In his groundbreaking tract, *What is a Nation*, Renan runs through all of the factors that could comprise a nation, such as a dynastic tradition, race, language, or religion but comes down to the idea that a nation is based on mutual consent: "a nation's existence is a daily plebiscite." The level to which national identity came about through multiple paths in no way weakened its potency. Even a synthetic nation, developed over the years of shared sacrifice, was still "a great solid unit" (Renan 1996/1882).

The third debate stems from the distinction between the organic or synthetic view of national identity. Must nationalism be tied to "blood" – a shared ethnic bond – or is it something that relies on shared ideas? These debates overlap considerably but are still useful to parse out. Should national identity be based on a distinct ethnic lineage or is it based on something else? A look around the world of self-defined nations demonstrates that while some share a cohesive ethnic identity, other nations are defined more broadly. These differences between so-called ethnic and civic nationalisms show up around the world but are also more of a continuum and a process.

None of the most ethnically defined modern nations began as such. France is often cited as a fairly homogenous country, and it clearly imagines itself this way despite the existence of much internal diversity. Yet French nationalism involved the progressive diffusion of culture from the Île-de-France and early state intervention deigned to stamp out local differences (Weber 1976). Like so many other countries, France experiences challenges to its national identity from new immigrant groups. Countries made up of many diverse cultural groups must define their national identity in some other way, often through a shared history or shared ideals. So-called territorial nationalisms describe how such nationalities must define themselves more multiculturally. The distinction between these two ideas of nations can even be applied within a single country as Breton (1988) does in distinguishing between French Canadian and English Canadian identities.

The confusion as to whether national identity is more organic or synthetic takes form in the experience of European Jewry, and in the quest for a Jewish homeland. From the Medieval period marked by quarantine, ghettoization, persecution and even massacre, most European Jews emerged in the 18th and 19th century eager to assimilate into the newly formed national societies. They would have considered themselves English, French, or German and they would accrue enough markers of belonging to these polities. Somebody like Theodore Herzl (1988/1896), however, emerged to point out that these were just mirages, that Jews could never truly belong: "every nation in whose midst Jews live is, either covertly or openly, anti-Semitic." Herzl argued that beyond being a religious minority, Jews were also an historic group with common characteristics and that all of these different peoples, speaking different languages in different countries, should band together and form a separate state. These writings sparked the national movement of Zionism and led to the formation of Israel which in itself incorporates an enormous variety of peoples.

The relation of the nation to the state stems from this question of whether national identity is more organic or synthetic. The ideal of nationalism is that humans are divided into nations and these nations claim loyalty on individual feeling and action. It makes the cultural political. But there are distinct ways of thinking about how this works in practice. One question is whether nations can themselves exist without states or without the prospect of states. Since nationalism is predicated on having self-determination, would a nation exist without this tenet. If so, the world would have been separated into nations well before the age of territorial states (Armstrong 1982). There would be such nations under imperial or feudal systems. Once a system of states was established, then they proved a welcome fit for the nations which were already extant.

If nations can exist apart from states, or apart from a system based around states, there is the philosophical and practical issue of how the two entities are conceived. There may be an emotional difference. The philosopher Georg Hegel made a clear distinction between the patriotism one could feel for a state from the affinity one could feel for the nation (Church 2011), one echoed by other scholars (see Huddy et al. 2021). Certainly, there is a functional difference between state and nation, but also a clear relationship. The state conducts the activities of governance; the nation entails the connection between peoples. An affinity between members of a nation must be nurtured and institutions are the means to accomplish this. While every nation can develop its own institutions, a state can provide or foster many of those essential ingredients – education, public propaganda, military service, government employment – that will endlessly promote the connections among members of a favored nation and relegate other nations to the sidelines.

The tightness of connection between state and nation is key to different types of nationalisms. Proponents of what would be considered "integral" nationalism believed that the nation should be sacrificed to the state and that more democratic institutions should be repudiated as obstacles to the nationalist project (Dahbour and Ishay 1995). This was a sentiment propounded forcefully by activists such as the 19th-century French nationalist Charles Marraus (Smith 1979). In such a view, nations were considered organically, based on race, "blood," and a language uncontaminated by hybridization (Gilbert 2008). This idea gained a great deal of popularity among fascist movements that constantly evoked the "nation" to justify state actions. To fascists such as the Italian Benito Mussolini and the Spanish Francisco Franco, the state was the instrument to actualize the national will towards greater expansion and glory. Both dictators valorized the dominant nation and forbade any expressions of minority national identity.

The opposite perspective sees the state as a separate entity; one that can host a dominant nation but one that can also accommodate other nations as well. This view also has a long history. For example, Lord Acton wrote that it was preferable for a state to include "various nationalities without oppressing them" (in Ozkirimli 2010, 26). More recently, philosophers have made convincing arguments for multinational states. Will Kymlicka (1996) has notably promoted a sense of common "citizenship" that can provide cohesion within multinational states, further uncoupling state from nation.

Material Underpinnings of Nationalism

These varied philosophies of nationalism would have had little purchase if the overall conditions favorable to nationalism did not evolve. Scholars and activists could articulate and encourage people's changing views but would have little influence without a change in the overall mentality. This was a fundamental transformation. As Kohn (1944, 9) pointed out:

> identification with the life and aspirations of uncounted millions whom we shall never know, with a territory which we shall never visit in its entirety is qualitatively different from the love of family or of home surroundings.

The evolution in people's understanding came through a greater sense of national identity – a horizontal allegiance of peoples towards one another. The material transformations allowed a sense of national identity to develop; the writings of various philosophers, politicians, and polemicists fashioned this sense of national identity into a nationalist ideology and determined how nationalism would be conveyed.

National identity is quite fluid. It is defined in separate ways, and it is conceived differently by members of the nation. Elites will not only vary amongst themselves, but their attitudes often diverge sharply from less privileged members of the nation. Higher levels of education, the opportunity to travel, friendships with people from various places all affect how one sees their nation or their group. And while a cosmopolitan attitude certainly can co-exist with a nationalist vision, it undoubtedly alters it (Bowden 2003).

The work of uncovering why national identity emerged in the first place has been left to 20th-century social science. The questions boil down to two. First, to what extent has national identity existed for an extended period of historical time, or has it been constructed to meet the demands of the moment? Second, what were the factors that led people to consider themselves part of a discrete nation rather than as part of some other cultural group?

One principal argument for the duration of national identity is that it is based on longstanding cultural differences. Proponents speak about proto-nations that existed before any nationalist movements were conceived. These "nations" carried no political ambition but would conceive of themselves as a distinct culture and people apart from their neighbors. Kedourie (1993) described this most succinctly in that "humanity is *naturally* divided into nations ... known by certain characteristics" (emphasis my own). This idea that nations preceded the modern era has been voiced by several other scholars (Smith 1996, Armstrong 1982, Nietschmann 1994) who point to the salience of cultural distinctions and the continued allure of a homeland.

This idea, often labeled primordialism, requires a bridge between local groups who would all know each other either first or second hand and the larger imperial polities which carved up much of the organized world. Feudal society represented another challenge entirely, as it involved interlocking and sometimes competing sets of personal loyalties between lords and vassals. Such ties hardly lent themselves to broad feelings of larger community. How would individuals feel part of a common culture that ranged beyond their immediate social and spatial surroundings and how could they maintain this culture under imperial or feudal political systems? Nobody suggests that these nations would cohere as they do today, but some assert that they developed a common culture. Smith (1986) argues that many nations – and he limits his argument to a few European nations – derived from an ethnic core. This would involve both a place and a people with a common name, myth of descent, and a coherent language. In successful nations, the core territory would expand to bring more places and peoples under its ambit. True national identity developed as the core expanded and assimilated similar peoples into a singular culture. The creation of the state system reified these national commonalities.

A diverging viewpoint sees national identity as something constructed or instrumental. People began to see themselves as part of a larger nation because circumstances made this choice more viable, practical, and useful. In these conceptions, there was not any pre-state national identity to speak of. Instead, the development of a modern national identity is tied up with the development of the modern state system – itself a novel form of political organization – and the growth of nationalism as an ideology.

The modern form of national identity shares a couple of attributes. First is that of widespread commonality. Nations would include a million or more people, so national identity exceeds the numbers of people any one individual can know. Within that large group of people, there has to be a degree of uniformity: language, tradition, history, and other attributes must cohere. Second is the presence of distinct boundaries. Even if one considers the prospect of larger scale cultural groupings in a pre-national world, cultural differences in these cases met mostly at a blur, as people with one set of characteristics blended into a group with another set of characteristics. Transitioning to a national identity meant that the blurs were replaced with sharp divisions separating the members of one nation from that of another.

Many scholars point to "modernization" as that cluster of changing material conditions which led to the emergence of national identity. Gellner (1983) argued that forces of economic modernization required a larger national sensibility with widespread literacy. The quickening of capitalism in the late middle ages promoted specialization and economic interaction. Society became much more differentiated, and institutions formalized activities that beforehand may have been accomplished through customs. Integration theorists like Deutsch (1966) pointed to a shift from face-to-face communication to the identification with common symbols of culture. Greenfeld (1992) argues that national consciousness replaced religion or caste as the primary identity. In her view, it was nationalism itself which led to a more modern outlook. Conflict theorists like Hechter (1977) and Gellner

(2003) examine how economic differentiation and the resulting demand for urban-industrial skills like literacy firmed up boundaries between groups.

Most theories of national development are bound up with written language and growing literacy. Anderson (1991) famously describes national identity as an "imagined" community. Since this identity is so much larger than any one person's acquaintance, it relies heavily on symbology and on areas of common ground. Both Anderson and Gellner have argued that the development of written national languages was a vital ingredient of national identity. For Gellner (2008), the thrust of society towards modernization and industrialization made it essential for most people to read and write. The need for schooling requires the acquisition of formal written language and means that people within a large group must now communicate (a point also argued by Deutsch 1966). It also separates language communities from one another. Anderson adds that the invention of the printing press enabled widespread publishing and spurred the written development of so-called vernacular languages. These then increased the sense of a common nationhood for many.

A larger national identity requires a unique symbology, so people become aware that they are part of this community. This could be induced through more horizontal networks, between peers and undirected by higher authorities. This flows from the idea of national identity as a "daily plebiscite" (Renan 1996). Distinct groups may see even common national symbols, such as a flag, in a different light depending on their own concerns. Witness how the American flag has been used in the service to all sorts of visions of what American nationalism really means (Americans and their Flag, 2016).

However, many scholars of nationalism argue that feelings of national identity need to be harnessed by an elite group. According to Smith (1979), it then falls to the intelligentsia, a group that is highly educated, with a more cosmopolitan but still nationalist outlook, and who are a part of this new cultural and economic currents. The intelligentsia are the vanguard who work to manipulate the symbols of national identity, adding to the national language, promoting the national heritage, and mobilizing the newly actualized national populace. Greenfeld (1992) also argues that nationalism was led by greater numbers of elites seeking to frame or reintegrate a new culture that forged stronger ties between the expanding elite and the general populace.

Of course, this is all much easier if the nation is tied to a state apparatus. The terms "state" and "nation" are often conflated, most notably regarding institutions such as the United Nations which comprises an assembly of political states. They are also combined, such as in the term "nation-state." But the two are quite separate. While "nation" describes a group based on cultural similarities, a "state" is a political entity. It is concerned with administering a territory and the people within that territory.

The development of states opens up enormous opportunities to either advocate for a national identity or to repress competing national identities. They have long used forms of persuasion and institutional unity as found in the educational system, the military, and the bureaucracy that brings the same vision of the nation from the center outward. Most such efforts have been nonviolent and subtle, though there are also cases where state-sponsored national identities are brutally imposed, as attempted by dictators like Benito Mussolini and Francisco Franco.

States and nations rely upon one another, and it is no accident that the growth of the modern state system in the 1600s was related to the development of national identities and nationalism. In fact, some would argue that it allowed several national identities to flourish as a means of legitimizing the newly formed and empowered states.

The emergence of modern states evolved over time. The forces of modernization articulated above were necessary but could have led to several outcomes as earlier advances had resulted in different political forms. According to Murphy (1996), the formation of the modern state system was spurred by an emerging sense that territory should be sovereign. In other words, authority over people and resources would be vested within a single territorial unit. As this became a governing principle adopted by rulers across Europe, it meant that empowering discrete entities, as opposed to having overlapping authorities (such as in classic feudalism) was much more attractive. This empowerment also came with greater integration within territories and more separation between territories. Events like the Treaty of Westphalia in 1648, and later independence and decolonization efforts, bought into this principle and led to the sort of system we have today.

The rise of the territorial state did not by itself augur the aspiration of nation-states. But nations as they were being articulated and formed in the 19th century expected to be contained within these territorial structures. The nation must have a home territory that was exclusively its own. During the 19th and 20th centuries, the nation-state ideal was perceived as a hallmark of a modern society. Especially in the 20th century, during the era of widespread decolonization, countries throughout the world gained independence under the banner of national development or nation-building. This was often difficult since so many of these nations were divided into many distinct and distrustful cultural groups.

The degree to which nations come closest to a congruence with state boundaries varies considerably, of course. Few absolute cases exist where there are at least some rival national elements within a state, a theme elaborated in Chapter 7. Even today, larger state-sponsored nationalisms cannot displace the various ethnic loyalties that persist within a given territory.

The Diffusion of Nationalism and the Rise of Anti-Colonial Nationalism

Early perspectives of nationalism and national identity were European based. Europe, along with the United States, was where the first self-defined nations and the first putative nation-states originated. However, it was not too long afterwards the idea of nationalism diffused into other continents, considering some of the local peculiarities. The New World colonies were affected first (Figure 2.1). Their distance from the European metropoles, whether France, Spain, or Britain, weakened their sense of shared belonging. While the United States emerged as the first newly defined nation, it was only a few decades before Mexico and most of the South American countries declared their own independence.

The creation of Spanish American colonies extended the Castilian nation to Central and South America. Over time, any sense of trans-hemispheric commonality was replaced by the differences between European Spain and the divergent creole societies that came to define themselves as nations (Lynch 1973). These were all movements led by the descendants of European settlers, with some buy-in from the various mestizo populations that sought to preserve or enhance their status in the new order, and which continued to marginalize the Indigenous and African descendants in their midst.

New World Europeans did not lead all independence movements. The most notable was the rise of Haiti as an independent nation-state (see Dixon 2008). As a French colony, St. Domingue was marked by extreme stratification, large scale slavery, and oppression of all non-white people. The imbalance between 32,000 whites and 500,000 Black slaves (with a small class of

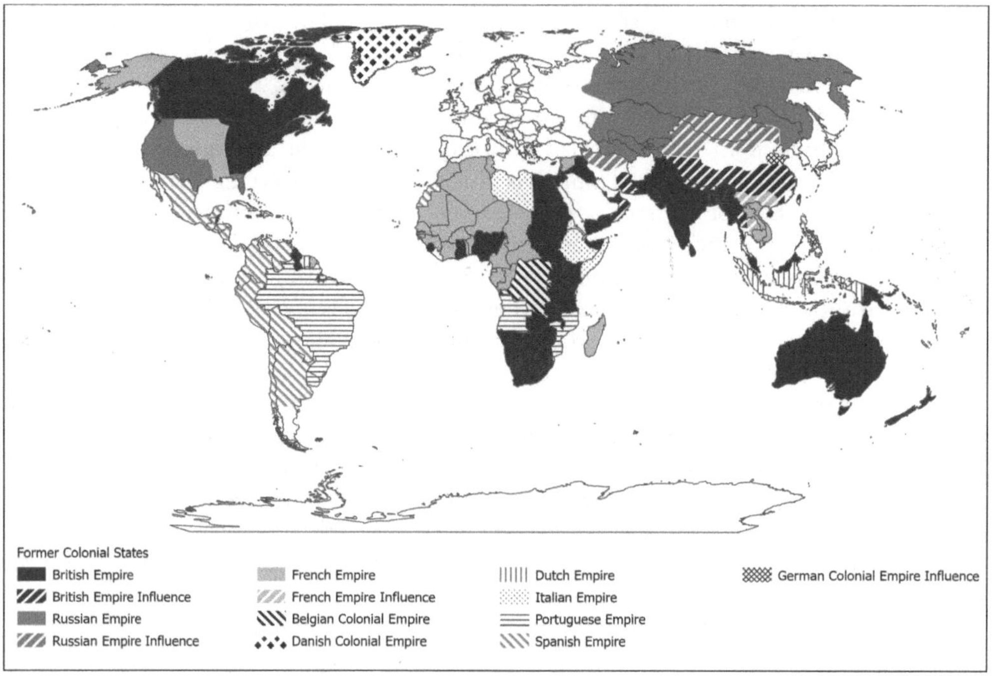

Figure 2.1 Former states colonized by European powers
Source: Stephanie Burkey

free Blacks and mixed-race residents in the middle) proved too much and Haiti became the first society to overthrow slavery and create a Black republic in 1804. The process of building a nation did not go smoothly as Haiti could not achieve political stability and economic prosperity. It was effectively ostracized from other countries worried about what its presence meant for their own societies; lack of interaction and trade hurt just when it was needed the most. There were also large internal conflicts between the small but well-off mixed-race population and the much poorer Black population.

Haiti was an exception in a 19th-century world order marked by both the increased national consciousness among some European societies and an expansion of European colonialism overseas, which completely disrupted the cultural and political geographies of these lands. It is hard to imagine what would have been the result absent so much European intervention; whether new national movements may have emerged organically. There were also broad differences in the nature of imperialism. New World colonization had already led to development of separate European-led societies which themselves sought independence. And most of North and Latin American was indeed independent by the end of the 19th century. European ventures into Asian societies varied from those where European powers sometimes colonized wholesale and sometimes exerted hegemonic influence over what were still ostensibly independent countries. One notable example of the latter was China during the 19th century, which was never formally colonized but was instead divided into economic spheres of influence dominated by various European powers (Parker 1994). Or India, divided between territories under direct British control and the Indian-governed Princely States. In many cases, existing cultural groupings were preserved, albeit under the heavy yoke of oppression.

It was in Africa and much of the Middle East that European imperialism obliterated existing cultural boundaries, carving up territory without any regard to what had been there before. During the Berlin conference of 1884, the African continent, which has been impacted by imperialism mostly in the coasts and still mostly independent of formal European control, was carved up by European powers.

The great land-based and overseas empires faded in the 20th century, as many colonized societies emerged as independent states. As part of this shift, well over 100 new polities were formed, changing the balance to a world where formal colonies were increasingly rare, though dependence still very much persisted. These new countries were often culturally quite diverse but nearly all spoke the language of nationalism. Yet at the same time the conditions were broadly quite different. In certain cases, such as in Rwanda, the Belgian colonialists had consciously divided the two main ethnic groups, making it difficult for cohesion to ensue after independence. Other new countries found themselves spanning dozens of separate language groups, different religions, and little sense of commonality. Other countries were carved from a larger culture region, such as all of the many states which emerged from the Arab world. And with each new country, colonialism had distorted the culture, the economy, and the politics.

The ideology of nationalism had been used to justify independence for each of the new colonies. Activists argued that European powers had no right to control nations half a world away. Yet at the same time, while nationalist ideals of self-determination were used to liberate countries, the new societies would continue along the same pathways, effectively replacing the faces without necessarily changing the system. Chatterjee (1993) argues that the new ex-colonies effectively created a Western state, adopting all of the institutions and manners of thought. Debates about whether a state should be secular, how it treats its own cultural diversity, or creating visions of modernity more appropriate to local standards – these could be seen as a different type of nationalism.

Circumstances have complicated the desire of many new states to chart a different nationalist path. Many new states emerged during a period of high tension between the Western bloc, led by the United States, and the Eastern-Communist bloc, led by the Soviet Union. While sympathies may have been with one hegemon or the other, the Cold War put new states in a difficult position and many quickly sought a path between the two main blocs. The Bandung Conference, held in Indonesia in 1955, was among the first attempts of ex-colonial countries to pursue a common strategy (Figure 2.2). It was attended by representatives of newly formed governments, anti-colonial, and national liberation movements, most in Africa and Asia (Bowden 2008). The participants sought cooperation between the countries, racial equality, and efforts to remain non-aligned between the two powers. The conference took place only a couple of years after the term "Third World" was coined (Wolf-Phillips 1987) as it proposed itself as neither the Western First World nor the Communist Second World. These initial efforts to remain non-aligned proved impossible in some cases as many of the attendees would later get sucked into the various proxy wars between the two global powers – the Vietnam War being a prime example.

A second issue was the relation of these new countries to various socialist and unification movements. Their initial goals were not just to create another European-style country built along a capitalist model, but they were not necessarily interested in replicating the Soviet model either. Some saw the need for societal transformation, particular inasmuch as the countries had been oppressed for so long. Thinkers like Leopold Senghor, a poet who became president of Senegal, spoke of negritude – the cultural values of the Black world and

Figure 2.2 Bandung Conference
Source: Picryl

the development of new ways of organizing society (Eckert 2013). It was also this view that prompted an African socialism which was a mostly African centered, economic model. Julius Nyerere of Tanzania was among the best exemplars of this approach and the African leader who best forged a true national identity within the state (Ibhawoh and Dibua 2003). Another push was in attempts to create pan-national movements that would transcend individual states. Gamal Nasser in Egypt sought to create a pan-Arab union, creating the United Arab Republic with Syria, a union that would fail shortly thereafter (Aburish 2004).

A third issue had to do with the extent to which these newly formed countries had to contend with so much internal diversity. Not long after independence, many new countries confronted cultural diversity and often major conflicts between separate groups, often occupying different regions. Many of the conflicts became causes of world sympathy, such as the Biafran secession movement out of Nigeria (which ultimately failed) and the attempted independence of East Pakistan (Bangladesh) from Pakistan (which succeeded). The term "nation-building" (Mylonas 2019) came to define a process of congruence between the boundaries of the nation and state, reflecting the ideology of nationalism. As discussed before, this goal is rarely achieved, but it was still held up as an objective for all new countries, both from outside and from within. This objective was thwarted by a lack of national unity in most cases, and violent secessionist movements in some. Economic underdevelopment and a lack of infrastructure also hindered attempts at nation-building since communication and other forms of interaction were not possible.

National Identity, Ethnic Identity, and Ethnonationalism

The idea of a nation, the construction of national identity, is the key basis in determining the ways in which different ethnic groups relate to each other within a society, and how ethnic groups are defined (Herb and Kaplan 2008). The advent of a nation-state ideal led to many different outcomes when it came to ethnic incorporation. While the ideal was in place, few countries managed to form a territory where the national group occupied exclusively the boundaries of the state and nowhere further. In fact, the geography of national groups was, and continues to be, messy.

Civic vs. Ethnic National Identity

There are many distinctions in how the nation is defined but the largest distinction is between the broad group of national identities described variously as "civic," "territorial," or "new" and those national identities described as "ethnic" or "old" (see Hayes 1931, Kohn 1944, Seton-Watson 2019). The principle behind a more civic national identity is that the national community is not defined by its cultural attributes, and particularly not considered on the basis of common heritage. The links between members of the nation can be ideological, as they are based around a particular set of ideals and aspirations, or by virtue of belonging to the territorial polity. Geographical membership in a state can be a powerful source of identification on its own. As Tamir (2019) argues, Kohn's explicit contrast of the civic nations of the West with the more ethnic nations of Eastern Europe implied a moral superiority to the former, especially in the midst of World War II. It also presupposed a sharper contrast than actually exists.

Many nations are constructed around ethnicity in some way, even if this requires a great deal of invention. There is also the civic requirement to try and harmonize the cultural diversity that exists within any polity and to incorporate newcomers, such as immigrants. Existing cultural diversity in several instances requires an effort to forge an identity which transcends the individualistic identities – a process sometimes described as "nation-building." Countries rising out of overseas colonies must negotiate several rival identities, which can threaten to fragment a society at its very beginning. Some countries, such as Tanzania, have been remarkably successful in developing a civic, national identity larger than the individual tribal identities. Other newly independent countries effectively displaced the existing Indigenous populations but opened themselves to mass immigration. Such an identity construction admits peoples from culturally diverse backgrounds. Upon entrance into the society, they are given the opportunity to enter the nation as well though the process can be slow, uneven, and fitful. Among the most notable of immigrant-receiving nations are the former English colonies of Canada, Australia, and the United States, but this model also applies to many South American societies. In these types of societies, new members are often expected to acculturate – to learn the cultural attributes of the nation, with language being the most prominent – but state-national identities constructed from existing cultural diversity may allow existing cultural differences to be politically recognized. We discuss further in Chapter 6 how multinational countries may function.

The idea of an ethnic nation is superficially simpler. Here the criterion for belonging to a nation is based on perceived cultural similarity. This normally assumes a common origin, so that all members of the nation descended from a sole source. Sometimes, however, new members can be admitted under the proviso that they "borrow" the culture and so become adopted as members of the national family – though this is arbitrary and by no means guaranteed. More often, the ethnic national identity is stretched to include groups on the periphery. The process of modern state expansion may not have initially had national commonality uppermost in mind. However, two trends made this more important. The rise of nationalist ideology necessitated that members of the nation could choose who represented and led them. In addition, national commonality came to be seen as a desirable feature of state integration. Especially during the 19th century, many European states became more self-consciously national and came to look at their dominant culture as the basis of the national culture.

The classic example of France showed what happened as a dominant culture emerged in the north and was moved to expand outward until it more or less reached the boundaries of the French state (Weber 1976). From the Île-de-France, the French nation, marked especially by the French language, came to encompass much of southern France where various dialects of Occitan were spoken. The growth of the French nation homogenized a language – and by extension much of a culture through various state institutions. Today, all of the Occitan languages are endangered (Crystal 1987). While much was lost, residual cultures remain. Notably the Breton language/culture – a Celtic offshoot – has stubbornly remained in the northwest peninsula (Figure 2.3). The Basque culture in southwest France remains as well though with less potency than the Basques in Spain. There are also border areas where French culture mingles with German culture.

Outside of Europe, a sense of ethnic commonality was also key. The Japanese state emerged from a feudal system, but national identity was based on long-time ties to the Japanese islands and to the distinct language and religious traditions that the denizens of these islands share. These distinctions have even been framed biologically, leading to a racialized sense of what the Japanese nation means (Murphy-Shigematsu 1993). Yet there are several peoples living in the Japanese islands who do not share "Japanese" traits. The Ainu population found in Hokkaido is Indigenous, distinct from the Japanese in appearance, and possessed of a separate language (which is spoken less and less) (Figure 2.4). The Japanese attempted to fold the Ainu into their ethnicity through intermarriage – a strategy rejected by Ainu activists (Siddle 2008). The burakumin and Korean minority are excluded from this national conception as well (Weiner and Chapman 2008; Neary 2008). And the Japanese who live outside the main islands, in Okinawa and to the south, are also only tenuously linked to the Japanese nation.

Ethnic nationalism follows from the premise that national identity and loyalty is predicated on cultural affinity. According to Walker Connor, nationalism is essentially ethnicity which has differentiated itself – is aware of its own particularity (Conversi 2004). Such a distinction also involves a political agenda in that members of the nation deserve

Figure 2.3 Linguistic landscape in Brittany, France
Source: Sebastià Giralt

to govern themselves. Under Connor's premise, non-ethnic types of nationalism simply cannot exist. Such an affiliation would be more aptly considered patriotism.

Connor (1993) also combined the two halves of "ethnic" and "nationalism" into a single term: "ethnonationalism." While clearly similar, ethnic nationalism and ethnonationalism carry distinct connotations. Ethnic nationalism clarifies how particular nations are constructed – arguing that cultural affinity is essential. In this regard,

Figure 2.4 Ainu peoples in Japan
Source: Torbenbrinker, CC BY-SA 3.0, https://creativecommons.org/licenses/by-sa/3.0, via Wikimedia Commons

Anthony Smith represents this idea that nations stem from existing ethnic roots fanned by ethnic entrepreneurs. Protean nations can exist before nationalism and as nationalism becomes dominant, these very proto-nations emerge as cultural communities with political goals. According to Connor, ethnonationalism relies completely on nationalist movements. It is completely ethnic in character, but it also emerges only with mass mobilization. Ethnonationalism involves the destruction of some putative nations in order to allow the growth of a single dominant nation. It is a mass movement that exists within the hearts and minds of its members. It is also an ideology which infects the very psychology of its members, requiring a suspension of rationality.

The Development of Majority-Based Ethnonationalism

Ethnonationalism has long mobilized self-determination for minority groups who wish to form their own country. We discuss these sorts of movements in more detail in Chapter 7. But we speak here of a movement associated with the majority population. Since Connor's piece, ethnonationalism has devolved from an affirmative identity, towards an explicit rejection of both civic nationalism and pluralism. Some forms of ethnonationalism are associated with majorities who worry about the dilution of their cultural and political influence. Ethnonationalist movements – often under the guise of populist parties which skew left or right economically – have crept out of the fringes and into the mainstream in countries around the world.

Majority-based ethnonationalism has been ingredient to the rise of nationalism itself as the boundaries around national identities are inscribed and re-inscribed. The best-known example came in the rise of 20th century anti-Semitism, brought to its horrific culmination in the Nazi Holocaust. For much of the 19th century, European Jews experienced greater levels of emancipation and freedom. Long set apart, they began to feel themselves as citizens of the countries they inhabited and members of the nation. Nowhere was this more evident than in post-unification Germany. With Jewish assimilation, Jews became solid members of German society, contributed to many of its scientific advances, and served eagerly in all aspects of German civic life. Anti-Semitism never went away, and as was true in many European countries, Jews were considered suspect members of the German community by many nationalists. The rise of the Nazi party in the 1920s and its takeover of Germany in the 1930s showed just how much Nazi German identity excluded and then persecuted Jews and other Germans.

After World War II, many (but not all) countries minimized blatant forms of majority-ethnonationalism, seeking to cast themselves in more multiethnic terms – even if actions and policies did not necessarily reflect these ideals. Recent years have seen something of a resurgence of majority-based ethnonationalism. The advent of "white nationalism" in the United States seeks to retain the American nation as a preserve of "whites," however this may be defined (Traub 2016). Other groups, while allowed to remain, are considered to make much more secondary contributions that must be sublimated to the national character of white America. The 2016 election of Donald Trump, while not explicitly self-defined as a white nationalist movement, was supported by many proponents of this philosophy. The election of the Law-and-Order party in Poland, while economically liberal, stems from a resentment of European liberalism and culturally alien newcomers (Traub 2016). The Brexit vote in the United Kingdom, which abrogated British membership in the European Union, was fueled by people who felt anxious about immigration and European encroachment on British values and control over their own borders (Kaufmann 2016). In addition, the strongest anti-immigrant sentiments occurred in those towns and villages that remained homogeneous, but where diversity was close by. The voting for ethnonationalist parties was highest near high immigrant areas, but not within these areas, and can be described as the "halo effect" (Rydgren and Ruth 2013).

Ethnonationalists fear that too much diversity will irretrievably transform the character of the nation. An empirical study by Inglehart and Norris (2016, 4) shows that support for populist parties "was strengthened by anti-immigrant attitudes, mistrust of global and national governance, support for authoritarian values, and left-right ideological self-placement." These traits fit into what the authors describe as a "cultural backlash" against a changing nation.

A couple of recent examples show how ethnonationalism can permeate political culture. In 2018, Brazil elected Jair Bolsonaro as president with the slogan "Brazil above all, God above everyone." In his appeal, Bolsonaro reframed what was meant by the Brazilian nation by excluding cultures considered insufficiently Brazilian, such as Indigenous peoples (Iamamoto et al. 2021). The government acted aggressively against pro-Black movements, arguing that this undermined the idea of a singular Brazilian identity. In addition to a reconfiguration of Brazilian nationalism, Bolsonaro professed support for an authoritarian form of politics, so much so that observers worried whether he would accept the results of an election which he lost. As it turned out, he did lose to former president Lula da Silva and, while the race was closely divided, Bolsonaro grudgingly agreed to step down, skipping the inauguration to flee to Orlando, Florida (Nicas and Spigariol 2023a). However, in January 2023, his supporters ransacked all government offices, claiming that the election of da Silva was illegitimate (Nicas and Spigariol 2023b).

Victor Orbán's Fidesz Party in Hungary has enjoyed more success in perpetuating its brand of ethnonationalism. Like the other movements, it is based around a fervent ethnonationalist ideology, hearkening to the tragedy of having lost much of the Hungarian nation after World War I (Toomey 2018). Fidesz also adheres to the notion of Hungary as a nation based on Christianity and a pre-Christian pagan past (Halmai 2018). However, unlike ethnonationalist examples in the United States and Brazil, Orbán maintains power over several election cycles. By manipulating the media, the constitution, and the electoral process, he ensures that his party almost never loses an election (Scheppele 2022).

Conclusion

The American and French revolutions set the stage for the growth of nationalist ideology which now encompasses the world as we know it today. Nationalist ideology is rooted in the idea that people are bound to the territory that they occupy, and that those people should have the right to self-determine their political and social lives as a united group. This idea spread across the world as the US and Western Europe became political hegemons, replicating the nation-state governance ideology through colonialism and soft power. Nationalist ideology is underpinned not by a key figure or set of characteristics utilized by every state, but by the overarching idea that nations do exist and should be protected.

Philosophical thought surrounding nationalism has roots in the 18th century and can be seen in examples like Jean-Jacques Rousseau, who wrote about the idea of "horizontal loyalty" or loyalty to the people as a whole rather than a governing body. Philosophical ideas revolving around nationalism fluctuate between views of the nation as cosmopolitan and universal, parochial and particularistic, natural and inevitable, constructed and manipulated, voluntary or involuntary. The two most widely understood and discussed views of nations tend to be understood as "civic," constructed through institutions that disseminate shared ideals and values; or ethnic, based on shared ancestry and cultural characteristics like ethnicity, language, and religion. Yet even these exist on a continuum and not everyone would agree that a certain case is an ethnic or a civic national example.

Members of strong nations, particularly ones who define themselves as "ethnic nations," tend to think of themselves as a part of an historical group which has existed for a prolonged period of time. In this view, states emerge as a complement to a natural and pre-modern national existence. Yet most scholars of nationalism view nations as a modern construction, born out of an impetus to construct homogenous populations within newly firm state boundaries. In this view, nations are instruments of the state, bound together by a shared sense of identity through education, literacy, and a common verbal and non-verbal communication system.

The question of how nations define themselves, and how they are parsed and studied is a murky one with no definitive answer. This non-answer lends itself to the very fuzzy and flexible nature of nations themselves. Nations are entities which exist beyond the state, yet the state and the nation are strongly linked. Nations need states to strengthen their bonds, and to fulfill their goals. States need national cohesion to be seen as legitimate and effective. It is no accident that these two entities both emerged in the mid-1600s, working in tandem to legitimate one another.

The dual concepts of nation and state which originated in Europe were exported around the world during colonialism and forced on peoples with differing views of territorial and group identities. Even as colonies began to fall, European national ideology stuck. In the Americas, European elites led most revolutionary charges which replicated European systems, while in Asia, Africa, and Oceania, power was transferred back to local hands, but under the justification of "national" self-determination. Thus, the nation-state system which originated in Europe was transplanted into more complicated and historically fraught situations where ideas of nation and state continue to be negotiated.

Over the course of the next several decades, nations and states became increasingly analogous in the geopolitical imagination. Nations were (and still are) expected to fit neatly into states, and states were expected to house unique nations. This articulates the powerful idea of the nation-state: a one-to-one territorial relationship between the nation and the state with total congruity. In practice, building the nation-state is a fraught endeavor, often leading to the state favoring of one nation over another, ethnonationalist development, and even genocide. A civic nation-state building may be less violent yet may not be able to achieve total homogenous group feeling among a cultural plurality. The concept of the nation-state was and continues to be an ideal rather than a reality.

Chapter 3

Territory, Networks, and Place

When we look at a map of the world, whether on a globe or on a flat map, we always see a patchwork of countries, color coded and clearly bounded. The only parts of the map that do not belong to a country are the major world oceans and Antarctica. By looking at this map, one might be led to believe that these patterns are stable, long-lasting, and cohesive. In reality they are none of these things. For one thing, a map is only a snapshot in time. What might seem like longstanding order can be disrupted at any moment, rendering an entirely new map in a matter of months. For another thing, each political map hides national nuance and nationalistic conflict in favor of the ideal of the nation-state, the idea that all states hold a cohesive national polity within their boundaries, priming a geographical imagination which defaults to a nation-state ideal. What is truly stable and long-lasting is the tendency to use this political map of color-coded countries to represent the division of the world into neat spheres of power and sovereignty, when the reality is far messier.

Two principles are reflected in the political map of the world. First, that the primary unit is the country or state. Units larger or smaller than that of state or country scale are rarely depicted, with an occasional exception for the map's "home" country, which may color code the primary subregions. Second, that the world is neatly ordered and clearly bordered. There is almost no ambiguity. Each colored patch indicates a clear space of authority. These maps are there for a reason. They are a form of propaganda, and an acknowledgment of just how essential territory is to the modern state. And because nationalism aspires to the self-determination of nations within the confines of a state, these maps also imply the geographical ordering of individual nations, people with common identity residing within a singular space and under the authority of a singular government.

We tend to think of this cartographic portrayal as axiomatic, but it is not. It exists as a moment in our geographical imagination. The only constant is our propensity to use maps to reflect power. The simplified universality of the political map obscures the complicated reality of power struggles, sub-national movements, sovereign self-government, and territorial projects. And just as the way we define and conceptualize state and national identity is complex, so is the case for territory.

In this chapter, we will discuss what is meant by territory and territoriality and how this relates to nationalism. We will then examine in detail the means by which nationalist movements (whether state based or insurgent) utilize territory: by impacting the discursive landscape, through demarcation and bounding, expansion, genocide or removal, and cooptation. Following this, we will examine the role of networks as a kind of challenge to these territorial spaces.

DOI: 10.4324/9780429435782-3

Territorial Strategies

Territory intersects with our modern state system in so many ways that it is tempting to consider it as a natural part of human existence. Indeed, territorial strategies in some forms have been observed for humans and animals alike. The initial ideas of territory came from the biological realm, where natural behavioral scientists like Ardrey (1967) explain how all animals seek to mark out their domains in order to control the resources within (Storey 2012). This can be achieved through a variety of signals. An animal urinating may also be marking its territory and letting others know not to trespass. According to Ardrey and others, less crude sorts of behavior can also extend to humans. People may mark their territory off with all sorts of symbols – graffiti, for example – as a form of spatial defense. While not necessarily endorsing the "instinctual" foundations for this behavior, researchers have shown the important relations between symbolic markers and spatial boundary-making. The symbols found in Los Angeles marking off gang territories, or the so-called "peace line" in Belfast that divided Catholics from Protestants are all ways that territory is demarcated.

This biological view of territory was brought into the political realm by Fredrich Ratzel, who derived an organic vision of the state and spoke about *lebensraum* or living space (W. Smith 1980). As an organism, a state (and the nation contained by that state) needed to expand in order to thrive, even if this meant the displacement of other peoples in the way. This idea could justify virtually anything, particularly the overseas colonial expansion of European powers in the 19th century, as new spaces for resources, farmland, and whatever else were greedily sought and acquired. Nazi Germany adopted Ratzel's ideas in its push to expand and eliminate existing populations, which contributed to their being widely discredited (Johnson 2018).

The more acceptable notions of territorial use come out of the socio-political view. These move away from the instinctual idea and towards territoriality as a strategy, used in a more nuanced way than the biological. As territory looks at the manipulation of space, it has been the role of two geographers to lay down some markers as to what this may entail. Jean Gottmann (1973, 1975) viewed territory as the means by which people controlled specific spaces. As such it is an expression of power. For political states primarily, this meant that one power, some supreme government, could have sovereignty over all of the land – people and resources – included within the territory. It also meant that territory could be spatially defended at the various boundaries, such as the physical impediments and checkpoints assembled at the US–Mexico border (Figure 3.1). While they gave up any independent claims, people within this controlled space could enjoy security knowing that they could live without worrying over outside marauders. This language of security is a large part of our present narrative and is intended to include the ability to check over and keep out unwanted outsiders. The United States has the Department of Homeland Security for this purpose, but every other country on earth at least attempts to control who comes in and out of their territory.

Bob Sack (1986) expanded on Gottman's ideas to develop a lengthy list of ways in which territoriality can be strategized. Sack begins with the premise that a spatial strategy is more efficient than other strategies. It simplifies the direction of human activity. It is easier to cordon off some rooms with "keep out" signs than to provide an accounting of items that cannot be touched. According to Sack, territoriality is the primary way to exert social power. At the level of the state, this can be used to classify particular areas, communicate the existence of this sovereign space, and make the entire organization seem

Figure 3.1 US Border Control
Source: Faungg's Photos

perfectly natural. In this regard, territoriality works with cartography in order to reify and make visible the state's power over its space (Harley 1989).

Territorial strategies have been employed in all human societies at some scale. However, the current use of territoriality in service to state sovereignty is fairly recent. As long as people have formed into groups there have been some forms of political identity. This is a shared social relationship involving some component of power (Tilly 2002, 61). Political identities would have appeared as soon as people organized into clans or tribes, where certain individuals were authorized to enact or enforce rules. Yet these sorts of political organizations did not manifest themselves in the same way (Ruggie 1993). In many earlier forms of political organization, people were defined based on kinship ties – being part of an extended family. Those people certainly occupied territory but the territory in no way defined them. Other forms of political organization may lead to clear claims over lands or migration routes, but these claims shifted with the seasons. Or there could be the creation of various lands over which some leader expressed control, but the control could be nonexclusive, with several rulers exercising power over the same parcels.

These sorts of systems highlight the fact that political organizations have always used space but not in the modern way. According to Ruggie (1993, 151), "the chief characteristic of the modern system of territorial rule is the consolidation of all parceled and personalized authority into one public realm." The territorialization of power bundles all government activities under one territorial umbrella, and the application of laws is thus administered by territory. The possibility of subdividing territory, which every modern state does, can allow for spatial variations in territorial control, the granting of certain rights by subdivided territory and so on, but the territorial dimension is key. This could be considered in terms of different language laws, as found in many countries with large linguistic minorities. A subdivided territory may allow for a separate set of language laws (say the addition of the minority language on street signs) than an adjacent subdivided territory (Laponce 1987).

How Nation Space is Territorialized through Spatial Identity

The value of territory to a nation is incalculable. All nations speak fondly of a homeland. Shelef (2020) argues that this homeland becomes the physical embodiment of the national identity. Homelands are rendered sacred in that they cannot be tampered with or bartered away. This notion also helps to consider what Kaplan (1994b, 586) describes as a "spatial identity": the land intrinsic to a nation's security and vitality. This land can be the actual space inhabited by the group, the environment and geography that now defines members of this group, and the historical legacy of a particular piece of land even if not currently occupied. So, this means that a group's homeland or its spatial identity can involve several different facets. In the case of Hungary, this might include land now controlled by ethnic Magyars, the physical landscape of the Carpathian Basin, and the territories where Hungarians held firm against invaders (White 2000). This also means that the spatial identity of a nation varies. For example, Catalan identity can include a more sweeping conception of a "greater Catalonia" which includes the province of Valencia, the Balearic Islands, the mini state of Andorra, and portions of France. It also can only include the more limited (and practical) notion of the four Catalan provinces in northeast Spain (Llobera 1990).

There are several ways in which the modern usage of territory impelled the organization and aspirations of nations. The simplest way is through transference. Nations may still exist without territory. However, nationalism is a political project that desires self-determination, and self-determination implies control over a territory. This makes territoriality vital to nations as well, as nations aspire to control a state territory. This also effectively distinguishes a nation from an ethnic group. Whereas an ethnic group is considered in relation to common cultural characteristics, a nation adds territorial ambition or control into the mix (White 2000). The territory fuses the nation together and anchors the group of people to a space. It intensifies the attachment of a people to the place and fosters a greater sense of national unity – at least for those people who belong to the nation – and provides a sense of timelessness (Penrose 2002, Moore 2016).

Despite this sense, nations did not immediately occupy the spaces they now claim. National leaders are eager to characterize their homeland as completely natural. Why wouldn't a group of people with a common history, a common culture, and a common destiny occupy a particular body of land? But the connections between nation and its territory were long, drawn-out affairs, still incomplete in many cases. Oftentimes, this was also reflected in semantic shifts (Greenfeld 1992). The development of the English nation followed a process whereby the geographical terms in usage changed. Prior to 1500, "county" was originally the unit in which someone resided. It had no specifically nationalist meaning. "County" then became synonymous with nation, changed in spelling to "country" and was also used with the term "patria" or homeland. The idea of the country was then seen as a sovereign political and territorial entity under the control of a monarch who in turn embodied the people.

How nations came to occupy territory also varied considerably. Several of the oldest nations in Western Europe developed from defined hearth regions. These core areas were also implicated in the development of states. In these early cases, the state would consolidate and expand into greater territory. With the development of nationalism, the dominant nation appropriated the state and state expansion, allowing for national expansion. The classic examples of this tendency include the French state emerging first in the Île-de-France regions surrounding Paris, then expanding outward. The "French" culture that was present in the Île-de-France, including the language itself, became the culture of the expanding French state (Weber 1976).

Other Western European nations developed out of multiple core areas. Germany most famously had no clear national starting point (Applegate 2008). There were multiple small states each with some facet of a larger German national identity, from free cities to landed kingdoms like Prussia. The consolidation of many of these small entities into the Holy Roman Empire for a millennium did little to foster a coherent German identity, though it did provide a template of Germany's spatial identity. This was true also of Italy, which had long held an inchoate national identity but did not fully form a state until the 1860s. Beforehand, the Italian peninsula was riven by multiple political entities, from some of the most advanced – like the Venetian Republic – to some of the more backward such as the Kingdom of the Two Sicilies. Beyond its political divisions, "Italy" was beset by strong linguistic variations, with several distinct dialects, and with low levels of economic integration. It was formed as a modern state only by merging all these differences into a whole (Figure 3.2). The phrase, "We have made Italy. Now we must

Figure 3.2 Italy 1859, before unification
Source: Nathan Hughes Hamilton

make Italians," credited (perhaps erroneously) to the stateman Massimo d'Azeglio expressed the herculean task that confronted Italy after it had gained its territory by 1870 (Hom 2013).

In many smaller nations that developed amidst larger empires, such as was true in Central and Eastern Europe, the sparks of national identity progressed from a small elite group of professionals, intelligentsia, and clergy and then to artisans and merchants and finally into the wider population (Hroch 1985). The initial impetus for the nation came out in small core regions, often in significant towns where they could more easily mobilize (Hroch 1985).

Beyond Europe itself, national identity could also be traced to a few core areas. Identities were often defined in contrast to the pervasive imperialism. In several instances, the development of a distinct overseas national identity stemmed from European colonists, who may then coerce indigenous people to follow along. This was most clear among the many Latin American nations that were forged from European descendants living in the larger colonial cities, but who co-opted aspects of their country's unique geography and demography to argue for a unique national identity. Indigenous groups were used as pawns in this national strategy and were soon sidelined after independence was achieved (Lynch 1973). The sense of difference, felt by generations of colonial residents far away from the metropole, was enough to provide the elements of a national identity that could then be galvanized by charismatic leaders like Simon Bolivar. In the case of Haiti, a sense of national identity and a resulting revolution came from the population of enslaved peoples (Dixon 2008).

For indigenous peoples affected by colonialism, the pathway to nationalism often entailed the revelation of a heritage that could then be re-fashioned. Burma, for example, had been based around defined territorial kingdoms prior to colonialism and nationalists would use this as a basis for the new spatial identity (Schendel 2008). Such nationalist movements hearkened back to a glorious pre-colonial past. In countries which had been carved out by European powers, irrespective of existing culture regions, the struggling new national identity would have to be pieced together from several cores. In a fairly successful case such as Tanzania, distinct cores from each of the separate ethnic groups needed to be woven together to create a single national tapestry. According to Otiso (2018) Julius Nyerere was able to reconcile these separate identities under an ideological rather than cultural umbrella, though he also promoted the entrenchment of Kiswahili as a national language. Other less successful examples, like Nigeria, never quite integrated the distinct spatial identities of the Hausa, Yoruba, and Igbo as well as other smaller groups. Nigeria has developed some of the basic structures of a national identity, but it is shallow and secondary to the pre-colonial spatial identities which exert far more impact on people's consciousness (Honey 1999).

If the correspondence between culture and politics, between nation and state, was simpler, then we could be looking at a world of coordinated spatial identities and non-overlapping homelands. But within every country there are competing visions. This can be based among different members of the same nation, as we can see more circumscribed, more exclusionary, and more open models of what a nation should be. They also exist because there is something like 5,000 nations forced into and between around 200 states (Herb 2018). Nearly every country in the world is multinational to some extent. Some have dozens of self-defined nations. In other cases, there are nations that straddle or span across several states, the Arab world being the most obvious example (Culcasi 2018).

National identities each carry a strong spatial expression. Inasmuch as the nation aspires to control a state and the dominant nation can use the state apparatus to promote its interest and quite often protect its dominance over minority nations, national space is territorialized. If some degree of autonomy is offered, a minority nation within a state may be allowed some control or autonomy over its own area. This is true in many sub-state regions, from the Cook Islands to Iraqi Kurdistan. Minority nations, without control over any government, have a far greater challenge and must employ their own territorial strategies.

Impacting the Discursive Landscape

There are a variety of ways in which a nation, as a cultural group, can utilize a fixed territory to advance its identity (Herb 1999, 2018, Moore 2016). "Discursive" entails all those items – both written and visual – that describe a nation's sense of itself. These include political rhetoric about the territory, maps, graffiti, or anything that writes the national identity of a group onto the landscape and in the process makes it their own. From this point of view, discursive signals can render a stronger connection between a group and a territory as the landscape both reads and writes the national culture (Moore 2016).

Discursive landscapes do not exist in a vacuum. A person simply coming into a space and looking at the various elements would not be aware of what these represented – would have no feeling of what it all meant – unless they themselves understood the context (Häkli 1999). In this way even some things supposedly natural are in fact quite contrived. The types of discursive items are part of Billig's (1995) banal nationalism inasmuch as the sorts of signals it sends may be fairly undramatic, marking how a land belongs to a nation without appearing overly triumphalist or contentious.

Edensor (2020, vii) provides an extensive list and discussion of these markers which could make up a discursive landscape. His book looks at the "dense spatial, material, performative, embodied, and representative expressions and experiences of national identity." Among the elements to be included are certain iconic rural landscapes that may be used to epitomize a specific national identity: the pampas of Argentina, Ireland's west coast, where lies the Irish-speaking Gaeltacht, the polders of the Dutch countryside, or a Moroccan oasis. There may also be especially representative sites which connote historic glory or symbolize its progress. In this category can be found the Arc de Triomphe, the Statue of Liberty, or the Taj Mahal for glory and something like the Sydney Opera House for progress. Edensor points out that the destruction of the World Trade Center towers in 2001 was especially searing because these had become symbols of American progress. Beyond the iconic, Edensor also shows the mundane landscape which subconsciously registers a national identity. Landscapes develop based on what the natural environment has to offer but forcing upon this the imprint of a particular plan. This leads to differences between national spaces. The long lots of Quebec, along the St. Lawrence River and some key tributaries (Figure 3.3), were indicative of the seigneurial system that established French North America (Kaplan 1989). The New England village landscape was yet another pattern established around a nucleus (Meinig 1979).

A discursive landscape can be benign, at least at first blush, and the aspects featured communicate a great deal about the features of the given national identity. Estonia, for instance, has staked much of its national identity on the rural beauty of the countryside. As relayed by Unwin (1999, 169), one tourist brochure describes "land and sea, jagged

Figure 3.3 Long lots along the St. Lawrence River
Source: © Foto: Ra Boe / Wikipedia

coastline and more than a thousand islands, high banks and picturesque dunes, stone-covered areas side by side with patches of fertile soil." Yet even in something as uncontroversial and "banal" as an exaltation of a country's rural beauty there is an element of demarcation – separation from another competing identity. In this case, Estonians have sought to distinguish themselves from the Soviet past of collectivization which had little regard for the integrity of the Estonian countryside (Unwin 1999).

Other times, the discursive landscape more forcefully shows a nation in competition with other national identities. Language represents one marker of identity and is readily displayed. Within a contested area, groups will fight over the language of place names, the name of the region, anything that can be written down. Even the language of graffiti is contested, indeed in some cases subversive.

Road signs are often a battleground of identities, particularly when a minority nationalism is attempting to assert its culture on the landscape. In Wales, long suffering from the imposition of the English language and culture over its territory, the Welsh Language Association advocated specifically for a change in the monolingual English language road signs spread across Wales in the 1960s. Road signs in this sense represented centuries of Anglophone domination and the marginalization of the Welsh nation. When appeals for bilingual Welsh and English signs were initially unsuccessful, a campaign of painting over monolingual signs, pulling down road signs, and smashing signs in large public weekend rallies was coordinated. Several groups, even other Welsh nationalist groups, sometimes felt the tactics went too far, but bilingual signs were implemented in Wales in 1983, with the Welsh language appearing above the English language (Jones and Merriman 2009).

The decision by nationalists to reinforce the discursive identity of "their" landscapes can be more violent than crossing out some place names on a sign. Removing evidence of a rival group's presence – changing the facts as they appear on the ground – is a well-worn strategy. Radovan Karadžić, the leader of the Bosnian Serbs, recounted in 1991 his plans to erase the Muslim presence by blowing up their foundation stones in order to remove any aspect of a Bosnian Muslim presence. The idea was to use this selective destruction to purify the Serb presence (Moore 2016).

Bounding and Regulating National Territory

Key to the development of the modern state system was the imposition of state territory. As Elden (2010) argues, territory is understood as a "bounded space" which requires investigation of both the space and boundaries. The modern state system involved conceptual changes in both. The control of the resources embedded in the land – and so control of the land itself – was a defining feature of all political-military competition. Political power derived from the possession of land. This was true in all developments, from the earliest city-states to the grandest empires.

The significance of boundaries followed from this competition for land. Two of the most famous of these – Hadrian's Wall and the Great Wall of China – were built to protect a flank of the Roman and Chinese Empires, respectively. The Emperor Hadrian personally traveled to the Roman colony of Britannia to oversee the construction of a 10-foot-thick, 15-foot-tall limestone structure with turrets at about every mile, at the narrowest part of Great Britain – some 80 miles across. This was less for defense and more as an imposed line of political demarcation and to regulate the flow of traffic (Wilkes 2004). Many Chinese dynasties engaged in the construction of the much grander Great Wall which ran 3,728 miles (6,000 km) and helped shield their civilization from northern intruders (Dalin 2004). These larger defenses were also accompanied by the development of walled fortresses around cities, castles, and other places needing protection. There had always been restrictions on mobility, but until the modern era much of this came through private means. Lords could dictate the activities of their peasants just as merchants had personal authority over the mobility of their workers.

The state system relied on a new conception of territory. This was more than simply bounded space; modern territory depended on legal aspects of sovereignty over a unit of land, but it also required advances in techniques, especially developments in surveying, cartography, and accurate measurements of space and time (Elden 2010). The state system expanded the notion of territory and endowed it with much greater precision. Strictly demarcated boundaries became essential to the territorial nature of the modern state system and to the restriction and regulation of mobility. The creation of passports, customs controls, the precise rendering of lines on a map were part and parcel of a new type of system. Interestingly, while boundaries between states were strengthened, internal boundaries (which did so much to impede the flow of commerce) were weakened or eliminated as regional rivals to state power were marginalized. Beyond this, states practiced their territorial authority through several interlocking aspects (Torpey 2000). Laws established which sorts of people could cross the boundary and when they might do so, new techniques were developed to identify every person in the world, and an entire administrative infrastructure was established for this purpose. The passport became the essential document for anybody who wished to pass from one state to another and then be able to return (Torpey 2000). It spread unevenly during the 19th century and became standard after World War I.

While the development of border security was effectively a state function, as state and nation became conflated in several instances, it also began to carry strong national ramifications. Since the state had the ability to decide which one of its residents would be allowed to travel outside its territory, some groups of residents, those deemed less loyal and considered security risks, would be denied opportunities to move inside and outside. This has been vividly displayed by the actions of the Chinese government against members of their Uighur minority, who often find their opportunities to cross boundaries severely restricted (Schenkkan and Linzer 2021). Also, passport controls allowed the exclusion of certain categories of people from moving inside, or if already in the country, being given the right to return back. The various Chinese Exclusion Acts of the late 1800s made it increasingly difficult for Chinese nationals to enter the United States but then later denied reentry to Chinese residents who went back to visit China, unless they had family or owned property in the United States (Torpey 2000, 99).

Passports and the general increase in identification and surveillance also played into the formalization of immigration. For most of history, there was no state authority restricting one's ability to enter and live in a territory. There were lots of other restrictions to be sure that hindered boundary crossings, but most states were not in a position to monitor this type of traffic. People entered countries in Latin America or North America under conditions of enforced bondage, labor agreements, or of their own free will. The various countries did nothing to stop the flow, allowing people to settle and work in their country, though making it difficult for them to become citizens. There were no undocumented immigrants without the existence of documents. Only in the later 19th century did countries begin to establish immigration protocols and it was primarily in the 20th century that immigration restrictions were put into place.

Even without formal protocols to govern immigration, most countries had a clear preference for certain types of immigrants based on ethnic/racial/religious categories. Immigration was a key factor in shaping a nation's character. The national identity within the United States was initially bound up with Anglo Protestantism (Kaufmann 1999). This identity was already exclusive as it considered Native Americans and African Americans as outside the national community in different respects (Trautsch 2015). This conception was rocked at first with the introduction of Catholics from Germany and Ireland – a situation that resulted in an extraordinary backlash and political movements to try and restrict them (Zeitz 2015). Subsequent migrations of people from various parts of Europe continued to alter the composition of the United States, making it less Protestant, less Christian, and less "white." Immigration policy became a key lever in shaping who would be allowed in. And the 1924 immigration legislation tried to shape the composition of immigrations to a way the United States had used to be, before the influx of peoples from eastern and southern Europe, as well as China and Japan (Archdeacon 1984).

Other countries followed much the same pathway, enacting policies to make sure that immigrants fit within their conception of national identity. Many European countries, divested of their far-flung colonies following World War II, fretted as former colonial subjects sought greater opportunities in the metropole. The potential flood of new immigrants interfered with the desire to preserve a mostly "white" national identity. New immigration policies codified this sentiment. For example, British identity was divided between the imperial Britishness of the Commonwealth, with its diversity of races, and the national Britishness of the UK itself, which sought to protect itself from this very diversity. The British Commonwealth Immigration Act of 1962 limited who could immigrate to the United Kingdom only to those of direct British descent, restricting most overseas immigrants to white immigrants (Triandafyllidou 2003). The anxiety with which the United Kingdom greets its non-white and non-Christian immigrants is exemplified in the so-called "Trojan Horse" affair in which Pakistani Muslims in Birmingham were accused of trying to take over the local schools and operate them according to Islamic law. While the plot itself was discredited, the allegations opened up the sort of anxieties about the increasing influence of Islam in Britain and the very real conflicts between separate conceptions of morality regarding subjects like LGTBQ rights (Shackle 2017).

Even the most liberal European countries have held a wary attitude towards immigration. France was an attractive country for migrants from other parts of Europe beginning in the late 19th century. By the 1960s it was also luring many people from former French colonies – many of them Muslim (Wike 2014). This led to concerns that these newcomers were changing the face of France (Ogden 1989). What had once been manifest in anti-Semitic and racist attitudes was now joined by a powerful anti-Muslim wave of antipathy (Figure 3.4) (Wike 2014). Germany introduced a guest worker program which allowed the labor of immigrants to come through the door, while restricting their ability to participate in the life of the nation (Martin 2002). This led many people from other countries, notably Turkey, to decide to stay in Germany. Yet partly because of the nature of their initial status as guest workers, Turks have functioned apart from the rest of German society, in many respects living in a parallel society (Mueller 2006).

Territories and bounding have required a shift in thinking, not only about state sovereignty, but also about national identity. As the world became more spatially articulated, it reflected the idea that peoples within a state territory should either share a national identity or negotiate the conditions of pluralism, something we discuss in more detail in Chapter 6.

Figure 3.4 Girl in headscarf, France
Source: Islamicus

National Expansion

The boundaries of nations are much blurrier than the strict demarcation of states. Members of a nation and the lands they inhabit do not always stop at a line but may blend into the lands of another nation. We have terms for this blur, such as Mexamerica, to describe borderlands where cultural interspersal occurs (Haverluk 1998). Secondly, nations have the tendency to cross political boundaries and be present within existing state territory. While a clear boundary separates France from Spain, both Catalonia and the Basque Country straddle this boundary. Third, as we established before, many nations and the spaces they inhabit exist within a given state. They may or may not predominate in official subdivisions, such as provinces or recognized regions. Fourth, the spaces occupied by nationals may be in dispute, argued by one group but denied by the other group. Quite often the official state "erases" the presence of any nationality in a space they control. They may apply a different name that diminishes their difference, or just prohibit the nomenclature altogether. Turkey has long described the Kurdish population as simply "mountain Turks" (Evered 2008). One recent example is how the government of Myanmar is displacing and deleting any mention of the Rohingya Muslim peoples (Figure 3.5) – considering them as not part of Myanmar's 135 official "national races" (Cheesman 2017).

Such disjunctures lead to several tendencies. Most flagrant is the decision by a country to simply invade or annex parts of another country, based on the claims that the national territory naturally belongs to the state. The classic 20th-century example was the continued expansion of Nazi Germany under claims of bringing German national territory

Figure 3.5 Rohingya refugee camp
Source: DFID – UK Department for International Development, CC BY-SA 2.0, https://creativecommons.org/licenses/by-sa/2.0, via Wikimedia Commons

under its control. In 1938, Austria and a portion of Czechoslovakia (Sudetenland) were annexed on the principle that state boundaries dividing nations were effectively illegitimate. Towards the end of the 20th century, Iraq's Saddam Hussein decided to cross state boundaries and swallow Kuwait with the claim that Kuwait was the 19th province of Iraq and had no independent national legitimacy. Iraq's attempted conquest precipitated a multinational war that Iraq lost decidedly.

The final example of this would be the decision by Russia's Vladimir Putin to invade Ukraine in February 2022. In truth, Russia's invasion had occurred eight years beforehand when Putin decided to annex the Crimean Peninsula from Ukraine by noting that most of Crimea's population was ethnic Russian. While that was technically true, and also true that Crimea had been gifted to Ukraine SSR by Khrushchev in the 1950s, it still represented an extraordinary territorial grab (Figure 3.6). For this action Russia faced several sanctions, but nothing which caused it to back down. Further incursions in 2014 included wresting control of the eastern Ukrainian regions of Donetsk and Luhansk through proxy paramilitary forces (Kramer 2022). The events in 2022 and 2023 involved an initial strike by Russia that threatened much of the Ukrainian territory followed by the surprising success of Ukrainian troops and their recapture of key regions (*New York Times* 2022).

Expanding state boundaries to unify the nation may be quite unwelcome, as was the case with Iraq and Kuwait. A guiding principle of the modern state system is that each country

Figure 3.6 Ukraine annexation map and breakaway regions
Source: Stephanie Burkey

should respect the boundaries of all other countries – a policy that has conferred a great deal of stability in the modern world system. Yet border imbalances exist and can create much consternation. Irredentism stems from the Italian word for "unredeemed" and refers to cases where people of one state claim an area in another state (often just across the border) whose residents feel a great national affiliation with them. This feeling is reciprocated, at least by some people within the target population. It was first used to describe Italian claims of Austrian and Swiss land wherein lived linguistically Italian residents (Chazan 1991). In the above case of Crimea, the regional government initiated a plebiscite asking the population whether it wished to join Russia or remain as part of Crimea. While there were allegations that the questioning in the referendum did not offer a full array of choices, the vast majority of respondents elected to become part of Russia. This provided the cover Russia sought to make the annexation, though it was done in the face of international condemnation (Kofman et al. 2017). Russia attempted this strategy again in September 2022, when it issued a referendum to residents of its captured Ukrainian territories as a means of cementing Russian legitimacy (Beaubian et al. 2022). This ruse was never accepted and in fact much of the affected territory was later recaptured by Ukrainian forces.

Irredentist movements are hated by the government that currently controls that region. Unlike a national movement within a country, it is more difficult to placate these with greater levels of autonomy. Since 1950, both India and Pakistan have claimed Jammu and Kashmir as part of their own national territory, based upon the fact that the area contains substantial Muslim and Hindu populations. This region, presently a state within India, was once part of a larger Dogra Kingdom that includes lands in Pakistan (Hewitt 2008). The majority of residents are Muslim with many concentrated in the Kashmir subregion on the border with Pakistan. These residents have sometimes expressed a desire for to be incorporated into Pakistan while others would prefer a country independent of both India and Pakistan (Hewitt 2008). India has responded in different ways, depending on the government in power. There have been special status agreements that allowed Kashmir some legislative and financial powers, but more often there has been India's attempt to snuff out any resistance. Most recently, in August 2019, the Hindu nationalist Indian government stripped Kashmir of any special status and just shut off everybody's internet connections for about six months (Slater 2019). Actions like this under conditions of irredentism do not involve just the state government and the sub-state region, but also the neighboring country. In this case, Pakistan is always involved, and the danger of Kashmir is that it could furnish the spark for a horrific war between two nuclear powers.

Genocide, Removal, or Forced Acculturation

Nationalism places the national state as the primary unit, above all others. The relationship of nationalism to territory also is found in the desire of some nationalists to purify the territory they control – in other words remove all traces of national difference. Sometimes this goes under the terrifying euphemism of "ethnic cleansing," but the effect is the same: devise some way to remove people perceived as non-nationals from the national territory. The three common strategies are through genocide, the mass murder of people considered different, forced removal of perceived non-nationals from the territory itself, and the compulsion that all non-nationals quickly adopt the cultural attributes of the majority. These three approaches are often used together, or differentially between distinct groups. They can also blend together as forced removal becomes a *de facto* genocide.

The term "ethnic cleansing" is fairly recent, based on its use by Serbians lamenting their persecution by ethnic Albanians in Kosovo province and coming into wider usage as a tactic employed in the 1990s Balkan Wars. The term "genocide" has a slightly longer provenance, dating back to 1944 as a way to describe acts "with intent to destroy, in whole or in part, a national, ethnic, racial or religious group, as such" (in Glassheim 2008, 436). The ideas behind both terms have been around for a long time. The Crusades were clearly an effort to clear out Muslims from the Holy Land. Efforts by rulers, or even the general population, to destroy members of another group goes back as far as history itself.

Yet it was in the late 19th century that nationalism and the aspirations of new nation-states would render it something more specific. As nation-states emerged from empires of old, the process of ethnic cleansing moved or removed cultural minorities who could survive an empire but were unwelcome in the new nations (Lieberman 2013). This was especially true in southeastern Europe and West Asia, where the decline of the multinational Ottoman Empire led to the succession of expectant national states, many of which sought to homogenize their territorial populations (Glassheim 2008). This process could be marked by forced expulsions of Turks, Greeks, or other minority groups from places they had formerly resided in peacefully for generations.

The early instances of persecution, even murder, on the basis of difference were often more haphazard. People would become enraged at members of a group, or rulers would exploit divisions for their own gain. New nationalisms sought ethnopolitical purity, and wanted everybody who was different to assimilate, leave, or be removed. The newly modernizing and bureaucratic state also became more militant and acted in the spirit of a national "people." In fact, Conversi (2006) argues that it was precisely this obsession with development along all dimensions – catching up with the West – that accompanied some of the more horrific cases of genocide.

It was the treatment of Armenians by the new Turkish state that represented an early case of systematic genocide. The Ottoman Empire, which in many instances had served as a model of multiculturalism with the millet system, now was reduced to a Turkish state based in Anatolia. This forced a major shift in Turkish national identity. Atatürk, modern Turkey's first leader, chose to cast the country as secular as opposed to Islamic and outlawed symbols of religious devotion. He also moved the capital from the cosmopolitan city of Istanbul (a city that had been head of the Ottoman and, as Constantinople, of the Byzantine Empires) to the Anatolian heartland of Ankara (Evered 2008). This redefinition increased intolerance of minorities threatening Turkey's new national identity. Greeks were by-and-large expelled, and other minorities were expected to assimilate into Turkish culture. The Armenian genocide occurred just before Atatürk's rise but while a group of Young Turks had gained effective control over the country. It began with the 1915 deportation of Armenians in Anatolia but was conducted in such a brutal fashion that it qualified as genocide, killing some 800,000 Armenians (Glassheim 2008). Other genocides over the 20th century, most notably the Holocaust, became increasingly deliberative – harnessing the bureaucratic power of the modern state with an atavistic fury at people perceived to threaten national integrity.

The removal of specific ethnic group, often done hastily with little preparation or warning, has been a common feature of states seeking national uniformity and as a way to open up space for their own nationals. While seemingly not as barbaric as genocide, it leads to the murder and deaths of all those people forced to uproot. The creation and bounding of new states from empires or colonies often precipitated movement. At times it was forced, and even if not explicitly forced, it was done by people who felt in mortal danger by staying in the new country. The partition of India and Pakistan in 1947 famously led to violent massacres against Hindus in what became Pakistan and Muslims in the new India. Hundreds of thousands fled, and as they tried to leave, "guerrillas burst out of the tall crops that lined the road and culled them like sheep" (Nisid Hajari, quoted in Dalrymple 2015). The violence of these forced removals would be replicated over again.

Forced removal does not just happen when new countries are born; it can be a pernicious strategy to accommodate national growth or to achieve other national objectives. Nazi Germany was again a poster child for this brutality as they forced residents of lands they conquered to move out and create *lebensraum* (to remember Ratzel's old term) for Germans. Poles, Russians, and many other people to the east were removed this way. (And after World War II, Germans who had lived under Germany's old borders were themselves compelled to leave ancestral lands.) Long before the Nazis, the United States engaged in a long-running eviction of American Indian tribes from their own ancestral lands. These forced removals, punctuated by acts of horrific violence, recurred through the 19th century as new territories were "cleansed" to make way for white settlement. The best-known example recounts what was known as the Trail of Tears, following the Indian Removal Act of 1830, in which 60,000 people were forced from their homes and lands in the southeastern US to a newly designated Indian Territory (now Oklahoma). Many Indians were killed in the process, by exposure to the elements and to disease (Silvern 2014).

Box 3.1 Indigenous Nationalism in Colombia

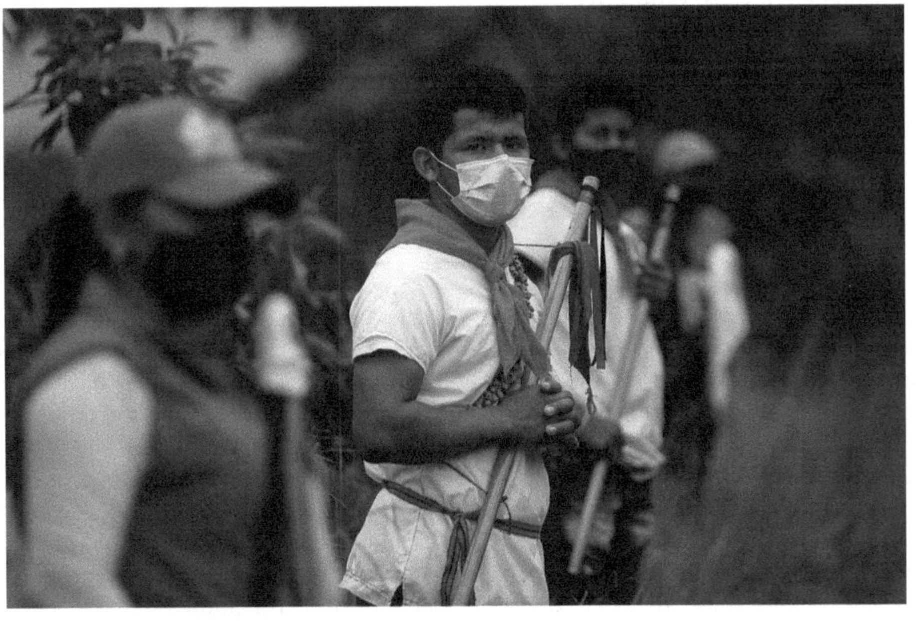

Figure 3B.1 Colombian Amazon Indigenous Guards
Source: Max Pixel

Indigenous populations were integral to the creation of a Colombian national identity (Jackson 1995). Descendants of Europeans formed the "Colombian" nation, mostly Spanish, but was heavily influenced by its indigenous roots (Rueda-Saiz 2017). The *indigenismo* movement began around 1920 and was led by non-indigenous intellectuals who supported Colombia's indigenous identity (Troyan 2007; Troyan 2008). These intellectuals dove into indigenous culture through ethnography and anthropological research, and this culture was then incorporated as an element of Colombian nationalism. What was initially a top-down movement eventually mobilized many indigenous groups who sought their own power within the state (Rueda-Saiz 2017; Troyan 2008).

One especially important realization was the inseparable nature of land and indigenous identity, especially the continuation of communal land use structures in rural and indigenous areas (Troyan 2008). During the time of Spanish colonial rule, large *resguardos* were granted to indigenous communities to enable their communal lifestyle. The newly independent Colombian government at first pushed for more homogeneous national identification and individualized land holding, so *resguardos* began to disappear. One law passed in 1890 protecting communal land holding was intentionally used as a means of forcing a "civilized" lifestyle on indigenous communities. As political *indigenismo* developed in the 1920s and 1930s, it included the perspective and authority of indigenous leaders (Rueda-Saiz 2017). The Cauca Regional Indigenous Council (CRIC), formed in the 1970s, sought the recovery of *resguardo* lands through detailed research. In the mid to late 1980s the Colombian government expanded *resguardo* land, allowing indigenous communities the ability to deal with foreign threats and drug cartels as they saw fit within the bounds of the Colombian constitution (Rueda-Saiz 2017).

Indigenous communities make up about 2% of the Colombian population but control about 25% of its national territory. They are empowered to maintain their own economic, social, and political systems (Troyan 2008). The historic influences of the state, missionaries, the Colombian left, and other indigenous groups on the definition of indigenous Colombian culture and identity has created a dynamic stage upon which indigenous communities seek to balance their ideas and goals (Jackson 1995; Troyan 2007).

One variation on this was the tactic of apartheid-era South Africa to remove Black African residents into "homelands" or "Bantustans" that the white-dominated South African government created (Figure 3.7). The term "apartheid" means apartness or segregation and its goal was to delegitimize the presence of non-whites from the largest and best portions of South African territory. The small, poor, homelands were the required residence of members of separate African tribes; they only could live in the rest of South Africa by sufferance (Paulin and Woodhouse 2008). Since it was only outside of the homelands where much work was available, what made apartheid such an evil system is that it created conditions whereby Black Africans were forced into a constant state of geographical uncertainty, always in danger of being displaced (Lemon 1976).

Compared to genocide and forced removal, compulsory acculturation of national minorities may seem benign, but its effects are far from it. It is important to contrast this with the voluntary acculturation immigrants around the world undergo in order to better

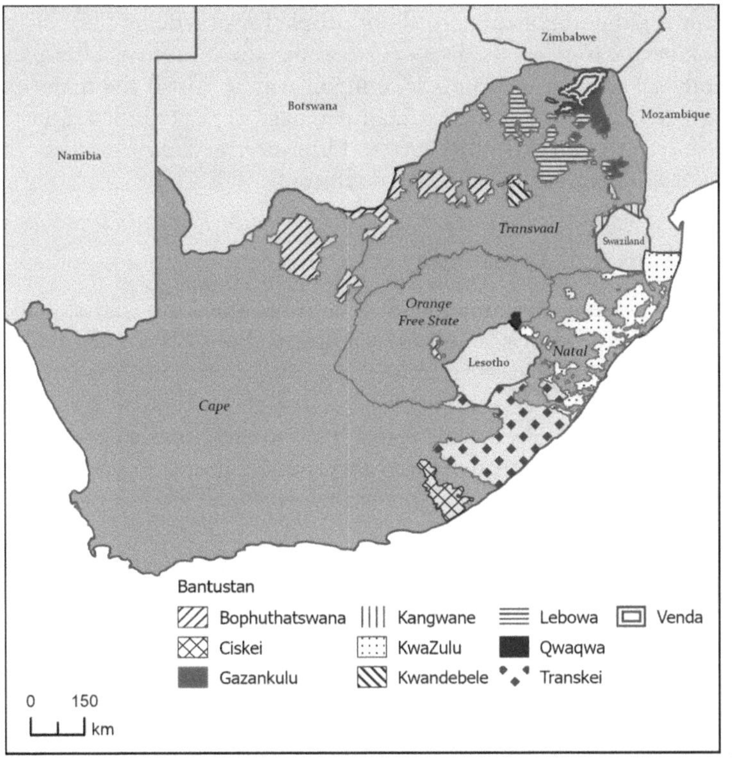

Figure 3.7 Map of apartheid-era Bantustans
Source: Stephanie Burkey

fit into their adopted societies. In moving to a new homeland, people decide it is in their best interest to learn the language and elements of the new culture. Most do not change their religion but will alter other aspects to the extent possible. Children of immigrants complete the process, retaining some aspects of their parent's culture while mostly adopting the new culture. Of course, things are quite different when a homeland is invaded, a boundary is redrawn, or a region retains its special national characteristics. In these instances, most residents have no desire to lose their culture. Yet nationalists of the majority may see this as an impediment to their vision of national unity and forcibly change the culture.

The nationalist movements of the 19th century took more aggressive turns in the early 20th century, and a few countries worked to promote a cultural uniformity across their territories by forcing minority groups to fully acculturate. The Spanish, under the control of Francisco Franco, followed on a path of forced acculturation. The non-Castilian speaking regions within the Spanish territory, notably Catalonia, Galicia, and the Basque country, were compelled to change their language of use. Throughout Spain, a single language, culture, and administrative system was enforced during the long Franco years. All regional bodies were abolished, all aspects of a separate culture were buried, and the military set up outposts everywhere to quash any regionalist sentiment through terror (Hannum 2016). Like Mussolini, Franco also encouraged Spaniards from Castilian regions to populate and dilute any regional distinctiveness (Conversi 2000).

Even without forcibly threatening to shoot people for practicing their distinct culture, a state has a number of tools at its disposal to render the minority culture less convenient and less desirable. Public education is a common way to instill the majority language in the younger generation, making the regional language appear outdated. Economic power also can translate into cultural power as acculturation becomes the best route for residents to move up or acquire desirable employment.

Transnationalism and Diasporas

The modern state system accommodates a national ideal of spatially fixed, bounded units, within which each enjoys a uniform culture and self-determination. Clearly this is an aspiration as opposed to reality; more examples contradict than uphold this ideal (Agnew 1994). Yet even within multinational states, the notion of spatially fixed entities prevails (Mikesell and Murphy 1991). States may attempt to create special autonomous regions, for instance, with their own governments and cultural privileges. Or, as we often see in borderland areas, the national spaces blur together, intermingling and creating a zone of confusion (Kaplan 2000).

While nationalism takes a defined space as its object, many national processes operate as flows. The idea of boundaries in any capacity are subverted by the interactions that occur across them. There is a global market for goods, capital, labor, culture, and just about anything. The recent experience with COVID-19 has demonstrated just how significant these interactions are, especially when many of the boundaries were shut down or tightened considerably. The term "globalization" reflects expansion of international networks though, like the term "modernization," it covers a great deal of ground.

One way in which networks challenge our fixed consideration of national space is through transnationalism. This is defined as a set of relations which cross political and cultural boundaries, in fact stretching across the world (Schiller et al. 1992; Vertovec 1999). Transnationalism takes many forms. A good economic example comes from multinational corporations. Such corporations work within multiple jurisdictions and maintain the flexibility and fluidity to circumvent the rules attached to these spaces (see Agnew 2017). Onerous environmental laws can be avoided by moving spaces; cheap labor is only a country away. Some political actors develop networks that are not bound to a place, but instead utilize networks of organized cells. Membership in terrorist organizations like Al Qaeda occupies various spaces but its loyalty is based on the flows between terrorist cells and the Al Qaeda leadership. This type of network subverts national identities (Juergensmeyer 2008).

Transnationalism may also challenge the territorial basis of national identity in more informal ways. Much traditional scholarship on immigration would examine how well immigrants adjust to their new context, changing themselves at the same time as they alter the places in which they settle. Acculturation and the placemaking impact of immigrants continue to be important aspects to study. Yet it is also complemented by an acceleration of movement and communication that allows immigrants more opportunities to maintain ties with their homeland. A discussion of transnationalism here involves the country of origin, the country of destination, and the linkage between the two. What is more, this specifically involves the ties between two localities as opposed to two countries. Translocalism describes the mobility of people and the exchanges between people as occurs between two places (Brickell and Datta 2011; Greiner and Sakdapolrak 2013). For instance, many highly skilled Asian Indian immigrants in Washington, DC feel unwelcome in the host society, and straddle the place they love and the place they come from (Chacko 2015). There may be several opportunities for a group to maintain ties with a home, from frequent visits, video chats, remittances, and even substantial investments.

One active form of transnationalism lies in the idea of diasporic populations. Diaspora originally referred to the Jewish peoples scattered across many countries where they were never allowed to assimilate. So, they maintained a distinct identity over generations, a desire to return to the Holy Land of Israel (when Israel existed as a concept), and some connections with one another. The term has become a convenient way to describe populations that continue to maintain a strong identification with a home country or with other members of the diaspora (Cohen 1996). The networked arrangements create an identity that does not fit comfortably in one space or another but exists via these linkages.

Several diasporic populations can be identified. One of the most prominent are the Chinese in Southeast Asia. The Chinese presence in Southeast Asia is of long duration, where they comprise a significant population in several countries, and are especially prominent in cities. While the situation differs by country, the relationship to the host societies is fairly uneasy. Many non-Chinese Southeast Asians view the Chinese as exploiters, because of their greater general prosperity and the historical role they played as "middlemen" under colonialism (Wang 1991). Some hosts also see them as disloyal, given their ties to the Chinese homeland and to each other (McCloud, 1995). It would be difficult to know where the Chinese diaspora fit under a standard spatial scheme. Their place in their homes is ambivalent and they have been kept at arm's length by native attitudes and have kept themselves at arm's length as a result of their own sense of distinctiveness. Strong transnational economic and social ties create a sense of belonging to a singular group with a shared set of experiences (Esman 1986).

Today, the expansion of virtual communication has altered transnational linkages between diasporas in fundamental ways. This began with easier access to international telephone calls (Vertovec 2004) and has accelerated with virtual meeting platforms, from Skype to FaceTime to Zoom, allowing the diaspora to remain in constant contact with family and friends (Brinkerhoff 2009). This shifts how diasporic populations frame their identity. Past groups coming to a country could get stuck in the past, remembering a home that was more of a nostalgic ideal, or a dystopian nightmare. New immigrants have current information about what is happening in their homelands. In the case of Polish immigrants to London, they have access to Polish internet, television, and radio. Children are folded into the national identity of their homeland (or their parent's homeland) as well as where they live at the moment (Kusek 2015). The blend of these two identities can create a hybrid identity, where diasporic populations forge a space between new and old identities (Hutnyk 2005).

Conclusion

Territoriality is the key difference between national groups and any other culture group. Territory can be thought of as land that has been demarcated by a group of people. This demarcation process lets those in the group identify themselves in the landscape and understand this place as home, while letting those on the outside know they are in a foreign land. Territoriality is key to national identity, whether state based or insurgent. Successful utilization of territorial strategies which clearly demarcate land colors our collective perception of global ordering.

Nations work to territorialize space by imbuing land with a sense of spatial identity. Binding a people to a place is done in a variety of ways. This can be done through discursive identity, which broadly encompasses visual and oral descriptions of a nation's sense of self. Nations utilize discourse to write, sing, storytell, and illustrate a national territory into existence. National territory is also firmed up by regulation and legislation surrounding borders. While borders are imagined political lines, they are made firmer by utilizing existing physical features like rivers and mountains, or made concrete by erecting walls and checkpoints, enforcing regulation of who and what crosses into national territory.

National groups without state backing often blur within and between borders. This can lead to national expansion, where states annex portions of other states which contain members of a kindred national group. These irredentist movements are geopolitically contentious, yet examples abound. Conversely, states containing national minorities may wish to purify the state territory in an attempt to achieve nation-state status. Such purification can result in "ethnic cleansing," genocide, forced removal, or forced acculturation.

Modern movements of people, ideas, and goods transcend many of the fixed borders which articulate nations and states. These de-territorialized networks account for a reconceptualization of territoriality while keeping the idea of homeland as central to national identity. Borderlands often blur distinct national homelands into new and unique entities, while transnationalism and diaspora exist with strong national identities apart from their original national homes. Globalization and modernization of communication technology allows such transborder networks to thrive, leading many to question the fixity of borders and territoriality itself in the 21st century. Whatever the reality, the idea of a territorial homeland remains central to the powerful concept of nations.

Chapter 4

Mapping and Symbols of Nationhood

Individuals and sub-groups that identify themselves with a nation rarely do so because of a keen understanding of national or ethnic origin. Rather, members of a national group identify themselves in national symbology that is deployed in obvious and subtle, "hot," and banal ways in their everyday lives. These symbols are national propaganda, made with the intent of promoting an idea of cohesive national identity, or with the intent of playing upon national identity to sell a product or an idea (Ichijo and Ranta 2016). The propaganda which has been selected by an elite group or state institution is then picked up and reproduced by private entities. These symbols include maps, flags, emblems, even names and landscapes. National symbology permeates the everyday, appearing in advertisements, books, music, art, fashion, textbooks, and architecture, just to name a few.

National symbology permeates our everyday lives in truly unremarkable and almost invisible ways. Postage stamps, money, flags, bumper stickers, and holidays have all become so commonplace and expected that they all but disappear into a landscape, covertly filling out the boundaries of the nation with symbols that are recognizable, comfortable, and unifying. Conversely, national symbology can be "hot" or highly contested when it appears in the popular sphere. Hot nationalism is usually expressed from internal national groups seeking independence. Graffiti, protest songs, separatist flags and symbology, or even political cartoons might all fall into the category of hot nationalism. Whether considered hot, banal, or even "warm," cultural products are in many cases considered by geographers as texts, social semiotic "readings" of popular culture that give us a sense of values, norms, and expectations within a group like a nation. These products both reflect and reinforce knowledge and power.

Differences between national groups are more frequently pointed out as our global populace becomes ever aware of our interconnectedness. Comparisons between national groups are often made by pointing to folk cultures. Dance, food, dress, and music from an ethnic past are used as qualifiers of national difference. However, these examples of national culture assume that a national character is the result of such cultural products and activities, and therefore immutable, stable, and fixed in an origin story of a pre-modern past. If that were true, then national identity would not be as contentious as it remains. You only have to look at the news to know this is not the predominant view of national groups. Rather, national identity continues to be stoked in our everyday lives, primarily through reiterations of the signs and symbols that denote the nation. So, it is not only historic folk cultures that delineate and maintain a sense of national identity among the masses, nor is it exclusively the national symbology produced by the state (Edensor 2004). Rather, it is the everyday consumption of various tokens that symbolize the nation and subliminally inform a sense of national identity.

DOI: 10.4324/9780429435782-4

Mapping, Cartography, and National Propaganda

The most evident symbol of a nation comes from the maps that surround us. A map is a powerful tool, showing borders, territory, important places, and drawn at a particular geographical scale. They are also tools of national propaganda. Instilling a sense of national territory begins in the education system and leans heavily on the utilization of maps. Maps are often taken for granted as accurate reflections of reality. They require mathematical accuracy to produce and are dependable for navigation. This has led to the faulty assumption that they are objective scientific artifacts (Pickles 1992). Maps are, by necessity, simplified distortions of reality (Monmonier 1994), produced by both public and private institutions with specific goals in specific social contexts. They are an illusion of reality, yet wholeheartedly believed by their observers (Wood and Fels 1992).

In the case of the political map used in classrooms, these maps show the world as static entities, uniformly colored to signal to the viewer the comprehensive power of the state therein. Existence on the map equals existence as a state and nation. But as any cartographer knows, artistic choices influence the communication of the data being displayed and can easily manipulate the perception of the viewer. Maps serve not just to communicate spatial data, but as an illustration of political mythology and ideology. They are able to be used in multiple ways to multiple ends, making them salient social documents (Kitchin and Dodge 2007). It is for this reason that maps are often deployed as propaganda in the service of nationalism.

Maps that propagandize a national goal affirm that cartography assumes the ideology of the creator. Propaganda maps effectively stir emotions and solicit brash responses, often by showing viewers something that they recognize, namely, "their" national territory under siege or threatened by some foreign other. This can be seen in wartime, for example in 1916 when *Life* magazine published a hypothetical map encouraging the US to enter the war (Figure 4.1), as well as in times of relative peace, such as in a map extolling the grandeur of the British Imperial Federation using strategic projection and color, not to mention the embellishing border illustration (Figure 4.2). National territory and the boundaries of it, so effectively drilled into the minds of a citizen, are seen as true, so the addition of propagandizing data is often taken at face value as well. Few people will stop to criticize the altered reality that a map portrays.

Maps are more than the sum of their parts. The very shape of the map is imbued with histories and memories, both public and personal. National maps are often categorized as "logo maps" (Anderson 1991), used as a popular reference to a meaning or idea rather than an articulation of spatial data. Logo maps are seen everywhere from clothing to television to newspapers and textbooks. The banality of a national border shown through logo maps is such that its invisibility is what makes it so powerful (Billig 1995). It is an instantly recognizable shape that conveys a cohesive and uniform entity, solidifying the idea of the nation as a similarly unified geographic truth (Monmonier 1994).

Political maps show space as a uniform entity divided by stark border lines. Maps aid in the assertion of political authority in faraway lands by reminding onlookers of the state's superiority in national and global geopolitics (Vujakovic 2002). Maps were transformed in the European Renaissance into the products that we know today (Harvey 1989). In this period, perspective and Cartesian rationality were added, applying a universal system of longitude and latitude to the whole earth. The standard elements of a map became recognizable and easily readable, such that they made a reliable and stable form of knowledge transfer (Latour 2011).

Figure 4.1 New Prussia map
Source: Persuasive Maps: PJ Mode Collection: 2160.01

The principle of national self-determination, propounded at the close of World War I, meant that each nation needed to prove its territorial extent to have a claim to self-rule. Because of their scientific respectability and artistic malleability, maps became a means of nation-building, promoting both internal cohesion and international recognition. Maps have been deployed in irredentist campaigns by Nazi Germany, Somalia, and Israel, in regional separatist movements in Basque country, Kurdistan, and First Nations in North America, through colonial projects and in service of nation-building projects in Jordan, South America, Finland, and the United States.

Mapping for Colonial Projects

Western colonial powers have proved adept at deploying maps in service of solidifying colonial power. Maps could make sense of faraway places, an enormous aid in colonial endeavors. The world was bounded into uniform entities that betrayed the heterogeneous nature of the places they represent. In such colonial cartography, indigenous people and local knowledge was wiped off the map, as a few examples demonstrate.

Imperial powers used maps not only to legitimize colonial projects, but also as instruments of the colonial endeavors themselves. European cartographers used blank space, lines, color, and symbols to advance territorial consolidation in 19th-century West Africa (Bassett 1994). With strategic placement of royal logos, cartouches, and imposed borders and boundaries,

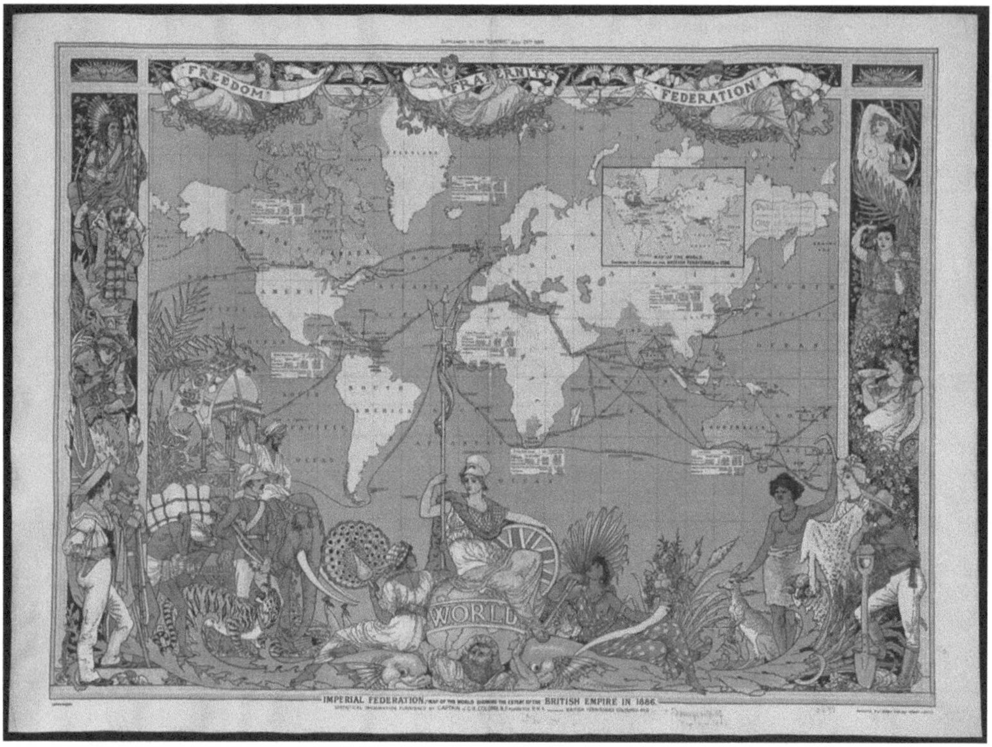

Figure 4.2 British colonial map
Source: Norman B. Leventhal Map Center

these maps legitimized and promoted the colonial effort. Simultaneously, the spatial detailing of resources, routes, villages, and physical geography aided in the practical invasion of the continent by facilitating military and trade movement and translating the landscape into a commercial possibility.

Mapping colonial territory was a haphazard process though. Late Enlightenment cartographers came up short in the ideal creation of a land survey that accurately represented the British-controlled Indian subcontinent (Edney 2009). Consequently, these poor maps hampered a cohesive sense of Indian national identity. This case shows how important visualizing the geography of the nation (through a Western lens) has been to the very idea of Western nationhood. Using these maps to communicate the idea of the nation impacted India's troubled nation-building approach long into the post-colonial period (Agrawal and Kumar 2017).

Western cartographic tradition prevailed over local indigenous knowledge in a drawn-out court case in 1980s Canada (Sparke 1998). Considered the only "legitimate" portrayal of space, Western maps and their users continue to enjoy authority over legal claims and property disputes. States that adopt and disseminate standardized maps have been more successful in nation-building, thanks in large part to the legitimacy that a Western map is still able to convey in a post-colonial world.

Imperial Siam was able to consolidate its power though maps, contested as it was between British and French colonial powers in the 19th century. Adaptation of Western mapping shored up Siam as a geographic body that could be interpreted by colonial powers, keeping occupation at bay (Winichakul 1997). Only by producing maps in a Western vein was the Imperial court able to reveal the existing geography of Siam to Western interests. In this process, however, Siam was also remade from the inside, as local knowledge and tribal customs were buried under a hegemonic Siamese monarchy. This marked the beginning of a Thai nation-building campaign.

Mapping the Nation

Political maps, like all maps, are trying to convince the reader of something (Pickles 1992). The claim of a group to a piece of land is only as strong as the representation that supports it (Peckham 2000). State-building is predicated on the construction of a cohesive national identity and, in order to "build" a stable and powerful state, the state's citizenry must positively associate the territory of the national state as the natural container for their group. The most effective and efficient communication tool for this message is a map. The viewer must believe that the map represents the shape, size, and relative global position of the nation's territory, and that this reflects the timeless order of things. If this goal is accomplished, the map has effectively proven the existence of the nation (Anderson 1991).

Herb (2004) describes the construction of a national homeland as a territorial process of differentiation, bonding, and scripting, a sequence made possible through the territorial imaginary brought about by maps. He describes this contentious process in post-war East and West Germany, two new states rendered from a cohesive German national identity. The contested nature of these new nation-building strategies resulted in several iterations of national maps from 1949 to 1970. First were maps from East Germany that ignored the new state border, portraying a unified German national identity despite the change that has occurred. West Germany chose to omit the Eastern state entirely. Next, East Germany built new ideological connections between their population and their territory. The textbooks of the time highlighted not the ethnic German bond to the land but the inherent bond between socialist character and Eastern Germany. West Germany then pivoted to producing maps that highlighted the negative effects of the partition more generally. None of these strategies resulted in a clear picture of German national identity during this time.

The modern Turkish state was built out of the collapsed Ottoman empire at the end of the First World War. In a region that contains many ethnic groups and faced with a brand-new territory to unify, Turkish elites sought to map the new nation. Turkish nation-building from the Cold War onward relied on propagandizing maps that stoked fear of bordering countries and suppressed minority identities (Batuman 2010). Popular justification for this fear of national outsiders came from Western rhetoric that posited the breakup of Turkey along ethnic lines (Peters 2006). The circulation of the Turkish logo-map developed cartographic awareness of a unified nation (Batuman 2010).

In 2002, Jordan, amidst a larger geopolitical identification with the Arab world (Culcasi 2012), and the recent end of Palestinian stewardship (Fruchter-Ronen 2013), sought to turn Jordanian attention inward, and build out a singular national identification with the state of Jordan. The campaign was called Jordan First. Culcasi (2016) shows how the use of map logo imagery in this campaign tied together larger regional struggles of hot political contentiousness with the more banal elements typical of nation-building (Billig 1995). The key image of this campaign is a stylized map (Figure 4.3), which weaves together the Jordanian flag, the shape of Jordan's borders, and a logo of three different colored hands clutching

Figure 4.3 Logo-map of Jordan
Source: Karen Culcasi

together. The Jordan First campaign has had mixed results among Jordan's diverse population and has been called a failed attempt at cohesive state-building (Schwedler 2012).

Mapping State-Led Irredentist Campaigns

Irredentist campaigns have leaned heavily on mapping propaganda as a way to communicate and solidify new or imagined national boundaries. Maps using suggestive colors, disproportionate symbols, and misleading lines are able to effectively show the casual viewer a world view based on a nationalist perspective. These cartographic techniques have justified territorial expansion but have also subverted colonial power.

The best-known example of maps being used for national propaganda is in Nazi Germany. The Nazi party famously began World War II by claiming portions of neighboring Polish land. The fact that ethnic Germans had settled in this territory, as well as the need for more *lebensraum*, or living space, validated the irredentist campaign. However, this information was expertly exaggerated and communicated to the public by German cartographers who extended propaganda maps from the earlier Weimar Republic (Herb 2002). Before the Nazis came to power, commissioned maps reimagined German territory to include pieces of neighboring states containing ethnic Germans, coupling post-war disgruntlement with geographic thought. This use of "suggestive mapping" planted the seed for the early Nazi irredentist campaigns, which were supported in Germany at the time. The Third Reich elaborated on the Weimar maps by adding disproportionate symbology that portrayed minority populations as large threats. These maps were displayed publicly, soliciting an emotional response from viewers. The Nazi use of maps as emotive propaganda highlights how effective maps can be in the dissemination of a national goal.

Similarly, Israel expanded its effective borders through portrayal of neighboring Palestinian land as ubiquitously Israeli. The Zionist effort featured propaganda maps from its inception, conceptualizing the idea of an Israeli state in popular imagination.

Figure 4.4 Blue Box for Israel
Source: דוד שי at Hebrew Wikipedia, CC BY-SA 3.0, https://creativecommons.org/licenses/by-sa/3.0, via Wikimedia Commons

Maps featuring the "land of Israel" were included on plenty of Zionist ephemera, such as stamps, posters, and pamphlets. Bar-Gal (2003) describes the "Blue Box Map" (Figure 4.4) included on donation boxes distributed among Palestine and the Diaspora. The Blue Box Map demarcated portions of Palestine that had been purchased by the Jewish National Fund, which became instrumental to the "land redemption" campaign of the Zionist movement.

The Zionist map of the Land of Israel was the result of a sustained effort to erase Arab ownership of the land, in an ideological sense (Benvenisti 2000). Sites and geographical features were changed from Arab to Hebrew words, while Palestinian villages were depopulated and erased (Kadman 2008). The campaign included widespread signposting, publishing of maps in books, newspapers, and magazines, and a physical re-shaping of the landscape. Notably, Zionist maps like those on the Blue Box were devoid of borders, leaving Israel territorially ambiguous (Benvenisti 2000). This runs counter to the logic that demands a national body be firmly established with territoriality achieved through cartographic bordering and othering (Paasi 1999b). However, Zionists were purposeful in their territorial ambiguity, leaving open the possibility of multiple statehood treaties, as well as the possibility of territorial expansion. Even after the establishment of an Israeli state, maps in textbooks remained ambiguous about the extent of the Israeli nation, and after additional territory was acquired in 1967, maps were quickly remade to reflect the Israeli occupation as legal and ordinary (Bar-Gal 1991).

On the flip side, Palestinians use the same geographical boundaries as Israel (with the exception of the Golan Heights) in the portrayal of their national homeland, referred to simply as Palestine (Wallach 2011). This depiction of Palestine as one geo-body, congruous with Israeli borders, rejects rather than reaffirms Israeli state control. Rather than considering these Israel/Palestine mirror image maps to be exclusive expressions of us vs. them, this demonstrates how maps are unique to individuals and groups.

Mapping Separatism

Maps depend not only on the context of their creation, but on the context of their use. Maps that have previously been used for colonial, irredentist, or nation-building projects might, over time, be laden with multiple meanings, or become devoid of national feeling at all. Some of the "hottest," or most emotional political cartography, comes from internal separatist movements. Maps depicting separatist aspirations are employed in much the same way as traditional state-led nation-building campaigns, put on posters, signage, clothing, flags, and other items. However, maps produced by national separatists are bottom-up, as part of a grassroots effort to realize a subversive political goal. Rather than appearing in state-sanctioned textbooks, map-logos of breakaway nations will appear in unofficial places, like graffiti and unsanctioned street art. These maps are concrete manifestations of a geopolitical hope. By putting the separatist idea into a map, the national territory is made real, even if not yet an administrative reality.

Maps of internal national entities, especially those with separatist tendencies, often show the national group as a sovereign nation-state, fitting into the Westphalian order with a legitimacy on par with any of the states that surround them. Paradoxically, this confirms the Western logic of the nation-state (Sparke 2005).

Many separatist groups employ the logo-map strategy to show the bounded unity of the nation, using new borders not internationally recognized (Anderson 1991). Basque separatists, for example, have expertly fashioned a logo-map depicting Basque country as one floating three-dimensional territory. Untethered to the jurisdictional reality of France or Spain, the two states that Basque country finds itself straddling, this map logo floats alone in space. The image has been reproduced on stickers, posters, patches, and tattoos, and is found all over Europe and beyond. Figure 4.5 is a photo of the map logo used in a sticker demanding Basque prisoners be repatriated. The photo was taken in Scotland, another internal nation vying for separatism. Mohr (2006) reminds us that hidden by this simple depiction of Basque territory lies a hidden administrative reality, including seven

Figure 4.5 Basque logo-map sticker
Source: Kathryn Hannum

provinces, two regions, part of a department, three nations, two states, and the European Union. The point is simplification. If Basque country is a unified nation, why should Spain persecute its political prisoners? This reordering and simplification of geopolitical space is a subversion of authority as well as a call to action.

The Kurds are an example of a national group that spans five different countries, Turkey, Syria, Iraq, Iran, and Armenia. Kurdish stakeholders in each of these countries have maintained some political autonomy within their respective territories, yet the fragmented nature of the Kurdish people remains a problem. The creation of a map of Greater Kurdistan challenges the borders, and therefore the national identities of all of these countries. Whereas Kurds are peripheral, or even invisible, to the national narratives in those five countries, the map of Greater Kurdistan places Kurds in the center (Kaya 2020). While some Kurds and international actors certainly use the map of Greater Kurdistan to rally action for political separatism, most Kurds use the map simply as a marker of recognition of their homeland as a unified entity (Kaya 2020). Either usage effectively makes the case that Kurds are a stateless nation.

Critical Cartography: Counter-Mapping New National Narratives

Geographers have shown a keen interest in the production and development of maps across societies both modern and ancient. Critical cartography and critical GIS seek to understand how maps (re)produce power relations and consolidate or subvert political norms (O'Sullivan 2006; Dodge, Kitchin and Perkins 2011). The previous sections have illustrated some of this research by geographers, focusing on the ways in which states have used maps in successful colonial and state-building projects. Traditional political maps in logos, textbooks, and media portray the world as ordered and neatly bounded. Critical cartography chips away at this cartographic façade and elucidates many of the structural issues present in maps and mapmaking, such as bias, privilege, and hegemony (Wood 2010; Harley 1989; Black 2000).

Critical cartography also highlights the fact that not all maps are read in the same manner. Map use is dynamic, as maps bring to light cultural memory, history, and both personal and group identity. Maps can work in affective ways (Aitken and Craine 2011), inciting emotional responses of love, loss, and pain. Maps are now being considered as cultural practices (Perkins 2004), narratives (Pearce 2008), and performance (Del Casino and Hannah 2005; Crampton 2009). What is more, as mapmaking has become more accessible, minority populations can be provided with the tools to make their own maps. New mapmakers have emerged seeking to produce alternative maps that both highlight and subvert power relations. The production of counter maps includes place names, boundaries, and natural features important to indigenous and sub-state nations. These counter maps re-insert marginalized people into the state-led national narrative or create a whole new narrative. This opportunity has been described as akin to the democratization of print media centuries ago (Peluso 1995).

Indigenous groups in Indonesia have taken part in an effort to claim some role in the mapping of forests and the resources. By claiming forest zones for indigenous community use as well as mapping non-state settlements, these Indonesians have added their voices to the state-led narrative of land and resource use and extraction. Likewise, on Turtle Island, Canada, indigenous counter-mapping projects claim local sovereignty from oil pipelines by reconceptualizing the state's conception of settled space (Hunt and Stevenson 2017).

States rely on an established idea of resource use for an imagined "collective" good. This collective is usually considered to be the majority nation. Minority stakeholders can easily be written out when states control national resources. In Bolivia, the government has worked to integrate indigeneity into the national consciousness (Postero 2007), incorporating very public indigenous rituals and indigenous rights legislation to support the move. However, an indigenous government does not mean universal indigenous corporation and concessions. When it came to concessions on resource decision making, attempts to give voice to all indigenous peoples in Bolivia ran into some obstacles. While the constitution ostensibly supports indigenous decision making, the indigenous state lays claim to the subsoil and the development of its resources (Marston 2019). In order to assert their claim over a region, indigenous groups from the lowland have taken to re-defining the territories they occupy to assert their claim over the space. Through participatory counter-mapping, some of these communities have successfully shed light on the contradictions embedded within Bolivian governance and have illustrated indigenous imaginaries of place in direct challenge to the state (Anthias 2019).

From first nations to national outsiders, counter-mapping can re-illustrate the very nature of national borders. Irregular migrants participating in mental and counter-mapping exercises show their overland and oversea journeys from a new, bottom-up perspective. These exercises show the reality of border crossings that are hidden by the state-produced macro narratives of immigration (Tazzioli 2015). Mental and cognitive mapping by irregular immigrants through Mexico effectively re-conceptualized borders not as fixed lines to be crossed, but as clusters of security apparatus reaching hundreds of miles into other state territories (Campos-Delgado 2018). These maps highlight that the new global reality of border security and border outsourcing involves power centers and interceptions far from boundary lines.

Nationalism in the Everyday: State Ephemera

Some of the most potent and yet banal nationalist symbolism comes from the state. An effective campaign makes people believe that the state has always been there. In part, this is achieved through symbolism. The design and deployment of timeless-looking symbols throughout the landscape gives the impression of state omnipresence, an understanding in the ubiquity of the state and the nation that composes it. Commonly used symbols include a coat of arms, national flag, motto, and national anthem, all of which are present on the landscape (Lowenthal 1994). Each time a citizen of the nation sees and recognizes one of these symbols, it unconsciously flags membership in the nation. These symbols may be simple, even disposable, but they are meaningful tools of nation-building.

The most widely circulated visual culture produced by the state appears on postage stamps and money. Geographers have taken to studying these artifacts as symbols denoting a statist conception of the nation. Both of these products convey a strong spatial association, as they are produced within and given exclusive jurisdiction over a state's territory. They fill out the borders of the nation, as they are easily acquired and mandatory in individual daily life. Beyond the utility, they contain a variety of images which convey the nation. Some convey an idealized national past, some show hopeful ambitions for the national future, while others convey abstract ideals about the intangible properties of some unique national character.

Money

Coinage containing political symbology has been around since the classical period, but nationally exclusive money is a fairly new concept, only gaining global traction in the 19th century after methods of mass production and extraction became viable (Helleiner 2018). This change coincided with a global rise in nationalism itself. The exclusion of currencies produced beyond the boundaries of the state solidified the preeminence of the nation and the concept of national sovereignty. Money – frequently used and often coveted – reinforces nation-state dominance on an everyday basis. The production and continued valuation of money fulfills a contractual obligation of the state and strengthens public faith in the institution of government. As currency maintains value and that faith in the state becomes stronger, so too do the nationalistic bonds that are symbolized by the images on the currency itself (Lauer 2008).

One of the first activities upon crossing international boundaries is the exchange of one currency for another. These foreign currencies pique much curiosity, as they look different and contain unfamiliar people and symbols. This enforces the otherness between national groups and makes firm the borders where one nation starts and another stops, where one nation's currency ceases to be valid and another becomes mandatory.

As territorial currency became the norm, so too did a widespread interest in domestic markets and a national economy (Helleiner 2018). Currency fluctuations also played into conceptions of national power and global standing within global economic order. This is what makes money a form of banal nationalism. People trust the value and validity of their national currency because they identify the people and images on them and rarely stop to consider the historical context or power implications of the coins and pieces of paper in their pockets (Gilbert 1999; Cohen 2018).

The state is not a separate entity beyond the society that the nation inhabits. Often, individuals and private institutions are responsible for the design and dissemination of national iconography that is considered "of the state," resulting in a more abstract conception of the state itself. Penrose (2011) investigates the process of bank note design in the devolved sub-state nation of Scotland. While Scotland is officially a region and not a state, the devolved powers of the United Kingdom afford Scotland the authority to print bank notes which portray Scottish national symbology. This corresponds not to the United Kingdom state, but to a sub-state Scottish nation. The money that is produced in "stateless" nations are nonetheless state artifacts that reinforce the *idea of a* state encompassing a nation, even if they are not produced by the state itself.

Today, fewer currencies are actually produced by the state. Regional currencies, adopted foreign currencies, and private and digital currencies are all available options for states to adopt. Regional currencies like the Euro or the African Franc, or the East Caribbean Dollar offer examples of states in Optimal Currency Areas that decided to relinquish control over national currency in favor of collective economic benefits (Wilson 2021). From these actions, scholars have offered an alternative "version" of national identity which deviates from the traditional and colonially imposed nationalism. Senegal and Cameroon, for example, have exhibited new facets of national identity which overlap state borders, forming a regional identity that is mirrored in the usage of a regional currency, the CFA franc (Ngwa and Akgün 2021).

Likewise, millions of Europeans find themselves able to identify strongly with their own nations and states but also with the European Union as it has built a "Common European Home" through dissolved border enforcement, common passports, and currency (European

Central Bank 2021). In the everyday, the Euro provides the most tangible link to the EU, in the coins and bills issued not by a state but by a regional coalition (Figure 4.6). The dual symbolism present in Euro coins reflects the dual identities that remain very present in the EU (Raento et al. 2004). On one side of the coin there is a generic logo-map of Europe, while the other reflects the national iconography of a specific country within the EU, reflecting the variety of national stories present within the coalition (Figure 4.6).

Digital money certainly erodes the symbolic power of paper and coins emblazoned with symbols of the state. That being said, the state is still ever present when thinking

Figure 4.6 Euro coins
Source: Kathryn Hannum

about and using traditional money, even digitally. Dollars, pounds, lyre, pesos, these are all regulated, valued, and set by the state. These monetary symbols and terms are in and of themselves banal reminders of the state's reach. Crypto currencies bring another dimension to the conversation, as suddenly the value unit is new and diverted from the state. However, even crypto currencies' value is measured against the value of the currencies set by the state, again reminding users of the state's presence and fiscal dominance. While most of our everyday transactions are done with the swipe of plastic or the click of a button, the architecture of digital platforms themselves draws boundaries around an imagined national community with whom economic exchange is simple and direct (Mihelj and Jiménez-Martínez 2021). Diasporas, expats, and travelers are likewise reminded of their outsider status when working with digital currencies. Sending money across borders, or engaging in foreign transactions requires conversion, taxes, and fees which are leveraged by the state and remind us of the power of the state and our place within it.

Stamps

Postage stamps also function as powerful illustrations of national ideals. Brunn (2011) has done considerable work showing how postage stamps are indicative of national perception. Post-colonial states, political and economic transitions, as well as social changes show up on stamps as "windows of the state," providing a glimpse into how the state sees itself at a given time both domestically and globally. Like money, stamps are expressions of sovereignty (Pierce 1996).

Stamps are issued and sold in themed bunches. Themes commonly include the following: history, such as founding events and documents, nature, including national animals and flowers, people, such as key political or cultural figures, science, and technology, like innovations or inventions, landscapes, like national parks or notable geologic formations, folk culture, such as instruments or crafts, economic drivers, like agricultural products or other exports, international events, like the Olympics or Nobel Prize winners. This list illustrates just a sliver of the themes present on stamps, all of which deepen the national story over time.

Stamps offer a blank canvas by which to convey ideas and ideals about the nation. Historical arcs are traced through the postage stamps of established states, but the very first stamp issuance of a new state requires extra attention, as it reveals national intent (Brunn 2000). New states have been analyzed through the postage stamps that they issue and are read as key texts in early nation-building efforts. In 1956, Israel's stamps were analyzed by Ribalow (1956) as indications of how the then nascent state was seeking to establish its national identity. Israeli stamps at that time looked to the biblical past by displaying images from the Old Testament. By portraying Old Testament scenes and Jewish celebrations in the Holy Land, Israel was engaging in a campaign of regional and national legitimacy.

The transition between the Soviet Union and Russia was likewise reflected in stamps (Brunn 2011). Soviet postage stamps contained a myriad of themes, including internationalist views and geopolitical stances. Russia, on the other hand, stuck to fewer themes in postage stamps issued in the early years of their new republic. Russian themes included nature, religion, and pre-Soviet Russian heritage, indicating an insular stance and an uncertain geopolitical position. Likewise, stamps issued in the host of new post-Soviet states in the 1990s were inward looking and stuck to basic symbols, like flags, coats of arms, government buildings, and natural landscapes (Brunn 2000).

Raento and Brunn (2005) examined almost 1,500 Finnish postage stamps from the 20th century. Themes were unsurprisingly centered around homogeneous ethnic Finnish identity, although the subject matter changed with current events. They point out that this myth of Finnish ethnic identity is key to Finnish national identity, and the ubiquity and acceptance of the concept of ethnic Finnishness is illustrated in the postage stamps used every day by Finns.

Mexican independence has been framed by stamps over the years. Stamps issued for the bicentennial celebrations in 2010 show continuity in the traditional story of Mexican independence, with changes in how the classic people and events are portrayed. The "heroes" of the Mexican revolution who used to be symbolized as static icons became actors in complex situations, signaling a change in their positioning as distant and untouchable to dynamic and human. Hoyo (2012) notes that the 2010 commemorative stamps offered a new view on independence itself, including a new distinction between the events of the revolution and the resulting independence, as well as a portrayal of internal ethnic and territorial differences. Both the constancy of the narrative of independence and the changes introduced in the new stamps align with political struggles around 2010. While actors critical of Mexican national history were present at the time of issuance, the mainline national narrative of the events being commemorated was kept intact.

Clothing and Textiles

At the 1984 Democratic National Convention, Jesse Jackson described the United States in terms of a bed dressing consisting of multiple stitched together pieces of fabric; "America is [...] like a quilt, many patches, many pieces, many colors, many sizes, all woven and held together by a common thread." This quote leads the recent Costume Institute exhibit at the Metropolitan Museum of Art in New York City and exemplifies how everyday objects are used to convey national identity. The common thread Jackson refers to is the national identity of America itself, while the many pieces of fabric constitute the many unique groups of people brought together in civic unity. The Met Costume Institute exhibit for 2021 was titled "In America: A Lexicon of Fashion," and showcased American born and American-based designers, cataloging how the clothing they have created has contributed to a uniquely "American" fashion aesthetic.

American fashion is best known for sportswear, and more recently streetwear. Ralph Lauren pioneered the sportswear aesthetic in the 1980s and incorporated traditional American symbology into multiple aspects of his brand. As a child of Belarusian immigrants, Lauren has said that to him America represents an ideal, which is reflected in his designs. His iconic flag sweater was debuted in 1988 (Figure 4.7) displaying the 1777 flag of a newly independent United States. The use of the American flag in fashion is now almost ubiquitous, as brands like Tommy Hilfiger, Gap, Old Navy, and others frequently use a stylized American flag as a logo. More recently, Lauren's flag design was reconceptualized to comment on the perceived unraveling of American ideals, first by Mexican-American designer Willy Chavarria, who wove a sweater featuring an upside-down flag with stars falling off (Figure 4.8), then by Black designer Tremaine Emory, who incorporated a woven African American Flag by artist David Hammonds into his sweater, signaling the interpretation and ownership of a powerful symbol among Black Americans (Figure 4.9).

In Wales, which claims to be the first industrial nation, materials like slate, coal, and wool are central to the national narrative and constitute physical aspects of a symbolic

Figure 4.7, 4.8 & 4.9 RL American flag sweater, WC American flag sweater & TE American flag sweater
Source: Kathryn Hannum

identity. As such, Welsh blankets and woven clothing have come to signify tangible pieces of a Welsh national past. These items are rich with symbology, containing both the wool produced in the water-powered valley mills and patterns unique to Welsh heritage (Thomas 2017). These traditional textiles have been revived by contemporary artists (Figure 4.10), while Welsh blankets are now sought-after pieces of home decor among fashionable Brits.

Both ethnic and civic elements commonly appear in fashions from around the world, signaling for the wearer national heritage and belonging to a particular culture. Fashion

Figure 4.10 Welsh blanket on loom
Source: image_less-ordinary

cultures encompass political attitudes and social moments. In China and Korea, authors Sun and Ha (2020) explore how visible symbols and subversive signals of national ideology appear in clothing. They list anti-Communism, collectivism, patriotism, and military ideology as invisible, even spiritual national elements which appear in clothing from multiple eras. Hwan (2010) conducted a study looking into clothing preferences among a large sample of Korean adults, and found that in addition to comfort, style, and individuality, national identity was a top consideration in what people choose to purchase and wear. These consumers sought national logos and Korean designers by name and tended towards a modern casual look which is being driven by Korean fashion brands like Hyein Seo, and SJYP. China, as an enormous retail market, serves to proliferate the expanding Korean, or K-fashion designs coming out of South Korea. A tension between anti-capitalist and socialist values emerges as the appetite for East Asian fashion grows. Sun and Ha argue that the design aesthetic present in K-fashion as well as what is chosen by Chinese consumers semiotically reflects the norms of national ideologies while catering to a globalized audience.

Popular Culture and National Identity

Our understanding of national identity is heavily influenced not only by state-sanctioned materials like money, stamps, or maps. Popular culture in the form of music, TV, comic books, literature, and movies buttresses our individual and societal conceptions of nations and geopolitical identities, and our collective ideas about geopolitics influences the production of popular culture. Dittmer and Bos (2019) succinctly describe geopolitics as an imaginary ranking of countries. Who has power, who is gaining power, and who is losing power? The power in question can be military, economic, or cultural, but all of these facets of "power" are influenced by and influence national identities. Geopolitical attitudes, that is, how people and groups see themselves within the flux of international politics, are structured by the informal and everyday consumption of media and popular culture in addition to the more formal structuring of nationalism done though state institutions.

Dittmer provides the example of being taught about the Cold War in elementary school while watching *Star Trek* at home. What *Star Trek* was portraying was just as powerful a geopolitical lesson of "us vs. them" as would be taught in school. In his *Star Trek* example, the US was the Federation of Planets, and the Soviet Union was the Klingons. This simple analogy of good vs. evil was absorbed by thousands of viewers through the 1960s and beyond, providing simple messaging for youngsters that may not yet be politically engaged. *Star Trek* remains a relevant pop culture example to this day as the show continues to be reincarnated in various series, movies, and books over the past half century, reflecting changing geopolitical realities and affecting new generations of viewers. While the original series was critical of nuclear war and US foreign policy (Sarantakes 2005), it was modeled on US–Soviet relations in a one-dimensional good vs. evil paradigm. Newer series are far more critical of American interests both domestically and internationally, incorporating critiques of race relations, foreign interventions, and conflicts. New story lines popped up to represent American allies in complicated and unflattering terms, while real-life "enemies" were given fictional incarnations that were sympathetic and multidimensional (Gunderman 2017). *Star Trek* went as far as to change historical outcomes in an exercise that urged viewers to critique current national events (Rhodes 2017). These changes reflect an evolution of *Star*

Trek's target audience, but also continue a tradition of critical commentary within science fiction. A show like *Star Trek*, or *The Handmaid's Tale*, or a movie like *Blade Runner*, gives viewers glimpses into alternate national futures which are paradigms of current national realities. While critical and alternative, they still operate on the assumption that national, or even international consciousness, is foundational to future (fictional) political happenings.

From television to cinema, action storylines are often set against real-life geopolitical events, reflecting popular fears and national moments. The litany of James Bond films of the late 20th century concentrated on Cold War East-West relations, entrenching a similarly "good versus evil" attitude in viewers as did the original *Star Trek*, albeit in a less critical manner. Post 9/11 Bond films shifted to portray new conflicts and new global nemeses. For instance, the 2002 Bond movie *Die Another Day* marks a shift, wherein African warlords, North Korea, and Middle Eastern terror groups replace the Soviet Union/Russia as the primary antagonist of action movies for the next decade (Dodds 2010). These plotlines reflected the prevailing anxiety among Western nations about security amidst the war on terror (Dodds 2008). Furthermore, these movies allow viewers to think about the world from a vantage point that disrupts nation-state logic. They imagine a trans-boundary world, while simultaneously using national identities as the framework for building such compelling narratives and characters.

Feminist scholars have long asserted that the idea of the nation and many of the concerns that surround it are highly gendered. In no other genre is gender as apparent as in action movies. Jeffords (1994) chronicles the national ideology of masculine individualism of Reagan era America as reflected in a variety of hyper-buff, hyper-violent movie protagonists from Dirty Harry to Rambo (the image of whom we have seen repurposed in Pro-Trump 2020 swag) (Figure 4.11).

Figure 4.11 Trump Rambo flag
Source: Photo by Kathryn Hannum

More recent action franchises continue this theme in more subtle ways. In the *Bourne* trilogy, women are the catalyst for action, but are not particularly mobile, stealthy, or in possession of action-oriented authority (Dodds 2011). These are established tropes among both academic and casual feminist viewers of cinema (Durante and Loftus 2016), yet the ways in which these tropes inform nationalist discourses remains covert. These viewings connect national narratives from movies to the classic national narratives of nations being personified as female figures that are imbued with qualities both maternal (giving, forgiving, wise, eternal), as well as virginal (unspoiled, lush, pure).

Comic books provide rich examples of popular geopolitics and the structuring of national narratives of security and gender. The personification of the United States in a character like Captain America helps to remind readers and filmgoers of key tenets of the "American" nation, namely innocence and progress (Dittmer 2012). Likewise, comic book heroes like Captain Britain and Captain Canuck are given characteristics that are supposed to be central to their respective nations, and story lines that reflect geopolitical developments. Even the relationships between these characters communicate national bonds and times of geopolitical tension, like those explored between Captain America and Captain Britain and the special relationship between the US and the UK (Dittmer 2009), or the juxtaposition of Canadian multiculturalism and America's melting pot ideology (Dittmer 2009). Likewise political cartoons satirize the state while at the same time playing on and reinforcing national tropes and identities (Dodds 2010).

Popular geopolitical narratives are seldom harmonious, reflecting the dissident national narratives that countries grapple with. India is a prime example of this, as arguably India's biggest cultural export, Bollywood films, portray a nation of Hindu ethnicity and cultural norms when that is not at all the national reality of Indian identity. Devasundaram (2018) argues that Bollywood films exoticize India for consumers with Western tastes. This reveals a dual national narrative in which Bollywood outwardly portrays a neatly packaged Indian nation-state identity for global consumers, whereas independently produced Indian films as well as Bollywood/Hollywood hybrids portray a more nuanced version of Indianness for a domestic audience (Devasundaram 2018). While all popular culture products reveal some political truth about the society that created them, they also position themselves in a world where portrayal of national identities matters.

Finally, video games are usually quite loaded with gendered, racial, and national consequences. These games, like much of pop culture, reflect a male perspective and are created for a male perspective (Harvey 2019). Single-player games remain the most popular type of video game (Klepek 2021), and often have the player assume a role in a society which mirrors some culture of the past, present, or future. National emblems are common in such games, be they real-world symbols of states or invented ones. The player simulates political situations and makes decisions for a collective or for their individual avatar. Games like these give signals to the players relating to traits possessed by their side or opposing sides. These often relate to the imagined national traits of real-world cultures. The information contained in these games subtly works to position nations in binary opposition, while enforcing static traits to certain groups, such as perpetual conflict, or high corruptibility (De Zamaróczy 2017). Newer games urge players to question the suitability of the group they have been assigned to, at which point players can defect and form new cooperatives or societies, mimicking real-world national disgruntlement and separatism.

Box 4.1 Symbolism in Flags

Figure 4B.1 Symbolism in flags – Puerto Rico
Source: Lorie Shaull

National identity feeds off potent symbols (Carrión 2006). Symbols are important for the state to maintain its connection with its population (Billig 1995). However, over time similar types of symbols can be used by those populations to establish strong ethnic identification apart from or in direct opposition to the existing state (Benwell 2021).

This ambiguity is aptly demonstrated with flags, a key symbol of national identity. On one hand, flags represent national values, maintaining national identification even when reality does not always match these values (Benwell 2021; Carrión 2006). However, flags also facilitate a group identification separate from state goals, turning citizen support to alternative political freedoms and agency.

In Puerto Rico, both the United States flag and the Puerto Rican flag evoke strong emotional responses related to questions of national position (Carrión 2006). Many view the Puerto Rican national flag as an ethnic and cultural symbol and US national flag as a national aspiration. However, Puerto Ricans disagree vehemently on the ultimate political position of the island and the use of the Puerto Rican flag stirs up controversy as to its symbolism and meaning (Carrión 2006).

With the weight of the United Kingdom's colonial history, the Union Flag is viewed through many prisms. In Northern Ireland, where citizens identify as British, Northern Irish, or Irish, there are strong positive emotional responses to the presentation of in-group flags, such as how self-identified "Britishers" will respond to the Union Flag. There are also negative emotional responses to out-group flags.

There are continued feelings of annoyance and uneasiness with the Union Flag for those that identify as Irish (Muldoon et al. 2020). In New Zealand, the Union Flag is supported, regardless of social or economic classification, so support for changing the flag is low (Osborne et al. 2016).

Appropriated versions of the Chilean national flag were common during protests in 2019 (Benwell 2021). Chilean flags were blacked out and shot full of bullet holes as a commentary on the flaws within the state. Citizens chose to use modified versions of the flag to depict a broken state system, meaning the use of the true Chilean flag is revered and reserved to represent Chile when the state is living up to citizen expectations.

Figure 4B.2 Symbolism in flags – Kosovo
Source: Michaela Gawrys

A recent example of state efforts to use flags as a unifying symbol to the state is in Kosovo, a country that declared its independence in 2008 and remains unrecognized by the UN. The Kosovo flag was created through a process of careful consideration where certain colors and symbols were excluded. This was meant to distinguish the new state of Kosovo from strong Albanian and Serbian citizen identification (Cosentino 2015). In Figure 4B.2, the Kosovo flag is displayed on public, government buildings while the citizens are carrying the Albanian flag. The use of the United States flag shows the maintained support of the United States and the role Americans played in ending the war in 1999. The state promotes a distinct Kosovar identity while its citizens maintain their ethnic identification as Albanian (Gawrys 2021).

Conclusion

Symbols of national identity are a powerful force binding individuals to a larger community. The state plays the principal role in propagandizing the nation, sometimes intentionally and sometimes unwittingly. The map is the single most salient indicator of a nation's extent, of its boundaries, and of its important landmarks. For members of a country, the map of their territory is instantly recognizable and serves as a visual stand-in. Where the majority nation suffuses the entire extent of the state, these maps are effectively national maps. They show where the people belonging to the nation live and the lands that they control. Disembodied maps, presented as logos and shown independently of their surroundings, show the nation as a political and cultural island. Expansionist maps, including neighboring territory not currently under state control, disclose a nation's goals and may be a harbinger of future aggression.

Maps can also be tools for those marginalized, unrepresented nations. The visual articulation of territory displays a nation's existence in a manner not otherwise possible. For advocates of national separatism, these maps are powerful aspirational symbols. They can make national goals seem legitimate and bring the nation into the world of ordered states. Maps have also been shown as useful ways to subvert the normal geopolitical system. Counter-mapping, often used by less powerful groups, displays places that would otherwise be ignored. It considers true alternatives to the boundaries as perpetuated and enforced by governments and offers alternate visions of where people live and interact.

The state enjoys a special capacity to produce symbols of nationalism that people use constantly. Currencies and stamps are the two most noteworthy examples of this. Economic life within a country is powered by money, and the money is minted by the government. Since modern times, every country has developed its own currency; to use another country's currency would indicate dependency. The currency in turn is populated with all kinds of nationalistic icons, representing the dominant narrative of the national story. In this instance, a change in currency is of enormous consequence, as with the proposed change in the US $20 bill from representing Andrew Jackson to Harriet Tubman. Recent moves by several countries to share a single currency, the Euro most notably, are a potent symbol of an economic and political union. Stamps are likewise ubiquitous in national life and here there are even greater opportunities to represent national triumphs and to introduce new themes into the people's consciousness. Big events are commemorated in stamps, but so are the mundane aspects of national life.

Popular culture includes a panoply of items, which is not directly produced by the state. Yet these items are significant symbols of national identity. Movies, comic strips, television shows, and video games show – often in allegorical terms – members of the nation who should be considered national heroes and national foes. Because they come from many sources, the messages are not always consistent, and can sometimes conflict with one another. In this regard, they represent the nature of national symbols as vital but often contradictory beacons of identity.

Chapter 5

Action, Performance, and Agency

How are national identity and nationalism experienced and reproduced by individuals and groups? A good clue is in how people engage in rituals and performances surrounding the nation. Performative rituals punctuate the yearly calendar in every culture, and carry a host of meanings, both personal and collective, both festive and somber. We can see the tangible evidence of performance and ritualistic celebration marking life milestones such as birthdays, weddings, and anniversaries, and the industries that service them. Similarly, we observe performances of mourning in the spectacle of funerals and burial services, newspaper obituaries, and roadside markers dedicated to victims of car accidents. These are obvious rituals that most of us experience frequently.

Some specific categories of performance surround national identity and are commonplace in most cultures. Ritual performances such as parades, holiday traditions, and sporting events reproduce the nation. Performance might also be used to subvert dominant narratives by generating alternative national sentiments. Much like the banal symbology that emblazons everyday ephemera, performative rituals are often the taken-for-granted actions, events, and ceremonies which bring national consciousness to life, and which, over time, can result in banal structures of national power.

Nationalistic performances are associated with core identities underlying national character. These fluctuate between public and private, social and political, conscious and unconscious, and scripted and unscripted. They involve agreed upon and mutually recognizable behaviors that carry meaning beyond the instrumental purpose of the act itself (Kertzer 1988). For example, without its symbolic meaning, a parade is no more than a collection of people moving from one end of a street to the other. But the costumes, pace, music, and dance create the spectacle laden with meaning (Quinn 2003). These events contribute to everyday nation making, energizing identity in different political and social contexts. Such performances, often tied to public events, simultaneously ritualize political policy and national identity. Moreover, because such events occur among groups, they require a commitment from the participants; they are perfect places for individuals to come together as a group and assert their ideology through bodily performance. Performances of nationalism and national identity occur often. Saying the Pledge of Allegiance while standing in an elementary school classroom or speaking in hushed tones at a national monument or museum exemplify the deep normality of national performance.

DOI: 10.4324/9780429435782-5

Performing National Tradition and Heritage

The selective preservation, popularization, and institutionalizing of tradition creates the substance of national performance (Hobsbawm and Ranger 2012). Traditional heritage is invented in the sense that select histories are shaped and preserved to serve specific narratives for groups (Kellner 1997). Performing traditional heritage replicates national mythology for current audiences and serves current aims. Ritualized performances establish and challenge the scope of group history, culture, and heritage (Schechner 2003). The ritualistic acts that go along with traditions are meant to cause emotional reactions in performers and viewers and serve as a model of group behavior. In the case of nationalistic performances, these replicate the importance of the nation in everyday ways, as well as remind a globalized world of the unique (albeit invented) heritage of national groups. The performance of national tradition embeds groups in a larger global tapestry of national imaginations. Through the ritualistic performance of national tradition, nations are able to establish themselves in specific places in specific times, drawing upon mythical stories from the past (Brosius and Polit 2020). Performing tradition resurrects stories of the past to serve modern objectives (Tivers 2002).

Performing group tradition expresses both internal and external nation-building. These performances can be organically generated by groups from within the state, as a means of activism, celebration, or separatism. The state may also encourage traditional performance. Vittorini (2006) chronicles this internal-external tension in India, where the state has carefully endorsed and curated a scene of artists and performers willing to reproduce the cohesive and singular Indian nation. At the same time, some local group traditions denote pluralistic Indian folklore and a more multifaceted Indian national identity. This tension speaks to the dualistic understanding of national culture, as either a top-down, and singular expression, or as a bottom-up and pluralistic expression of nationhood. In India, many consider top-down, "elite" culture to be classic, whereas bottom-up culture is considered backwards. The multiplicity of national expression in India parallels a common political line "unity in diversity" which acknowledges the vast cultural diversity within India, while promoting the importance of the unified national culture.

Folklore is a common way that national traditions are expressed (Kapchan 1995). Folk culture – made up of street art, clothing, crafts, dance, music, and painting – draws on national mythology. It affords a sense of cultural stability, a touchstone for an imagined cultural past, and provides an everyday simplicity open to all, especially those who may be excluded from so-called high culture (Bronner 2002).

People think that they are passing folk traditions from generation to generation within families. Yet ethnicity and national identity are linked, and a select few ethnic identities are given a privileged position within a nation. In this manner, an ethnic tradition operates as the default model of national identity. This type of ethnonationalist framing can be covert, as is common in centering majority ethnic customs during national festivities, like Christian traditions during December in Europe and the Americas. It can also express xenophobia overtly, as in rhetoric which describes the mere presence of immigrants and minorities as "corroding social values" or "increasing criminality" (Rydgren 2003).

Hegemonic ethnicity is made and remade through everyday practices of performance, storytelling, and embodiment. These group values are enshrined in national mythology and played out in everyday life. Boundaries between national and ethnic heritage are constantly blurred and made clear on purpose, maintaining the hegemonic status of some groups while othering ethnic minorities (Sasson-Levy 2013). The centering of ethnic stories in national lore is often aided by an archetype, or a character that can be deployed during celebrations, referenced in art, and used in the performance of national stories.

Festival Celebrations

As celebrations of group belonging, festivals are forms of intangible cultural heritage and an affirmation of national identity (Nurse 1999; Del Barrio et al. 2012). A community also celebrates the values and ideology which has led them to single out a particular event for celebrating (Getz 2010). The function of a festival or any assembly is to physically capture commonly held notions of community in a single place and time for the whole community to see (Bennett and Woodward 2016). Drawing upon historical and often mythologized elements of national imagination, celebrations bring history and mythology to the present, and allow societies to actualize present needs using old stories (Jeong and Santos 2004).

The relationship between the state and the nation provides a groundwork for national holidays and institutionalized celebrations. These celebrations constitute an action that is structured around school being let out, shops being closed, bank holidays, and symbols being deployed at the expense of municipalities and regions. The performative rituals that are associated with these state holidays are often dramatic and always express official narratives of the nation. Many states have one day which celebrates a singular national story, often invoking mythic imagery and an uncritical narrative. These may be celebratory, as with the American Independence Day on the fourth of July, complete with barbeque cookouts, picnics, waterfront games, and festive fireworks. They may also be somber, like the British Remembrance Day, which is the only national day honored by all nations within the Commonwealth (Harrison 2013). It is marked by a two-minute silence and the ubiquitous poppy flower to honor fallen soldiers (Figure 5.1). While varying in

Figure 5.1 Canadian Remembrance Day
Source: GoToVan

source material, tone, and tradition, they are joined to represent and recreate a united national narrative.

By looking at key features surrounding the tradition and ritual of national holidays, Elgenius (2018) argues that as a rule, "all national days honor founding myths and official national narratives that glorify nationhood in terms of oneness. National days are designed in such a way that they perpetuate notions of sameness and of shared experience" (p. 95), although to varying degrees of success. The common elements of national days primarily include the object of the celebration. What is being honored is usually a historic event surrounding the founding, genesis, or safeguarding of the state. Second, whether ritualistic participation is limited to the military, government, or elites, or can be shared by the general public. Third, the strength of national symbols deployed as a "shortcut to nationhood." Finally, the degree of state sponsorship surrounding these events.

Tourism and Performance

National tradition is often performed for tourists, bottling selected historical, social, and cultural place identities in a short amount of time (Frew and White 2011). Because tourism is a major economic engine in many places, there is funding for putting on elaborate performances which highlight the uniqueness of national stories and peoples (Zhang, Decosta, and McKercher 2015). Celebrations tied to tourism are seldom conflated with nation-building, since the realms of tourism and politics are envisioned as separate spheres. However, cultural festivals and national identity are strongly linked in that they celebrate real or imagined notions of the past in order to serve a modern sense of belonging (Yang et al. 2020; Nelson 2021).

These performances of national heritage are unavoidably tinged by outsider expectations, especially when performed for tourists. This contributes to an altered sense of tradition, which in turn produces a mythologized and westernized sense of national identity. This has been found to be the case in many post-colonial states, where national identity itself has been imposed upon local groups. A post-colonial identity is then reproduced for outside groups from the very places that invented and exported the idea of the nation.

Zhang et al. (2015) show how post-colonial Hong Kong fashioned a national identity for themselves using colonial narratives as the difference between them and other Chinese cities. At the heart of Hong Kong's quest for unique identity is the ongoing tension between their domestic and international status (Chun 1996). While Hong Kong is administratively under Chinese rule, their national mythology tells another story (Fung 2001). Utilizing several myths from Hong Kong instead of mainland Chinese lore, performances geared toward tourist consumption have crafted Hong Kong as a destination, and a people, apart from China.

Heritage performances and historical re-enactment constitute a growing body of scholarly interest in the global heritage tourist industry. This global heritage tourist industry has been fueled by the United Nations UNESCO heritage program (Brosius and Polit 2020), which seeks to codify types of heritage, one of which is being intangible, and reify them for commodification of local industries. The language of UNESCO and other tourism memorial entrepreneurs (Alderman 2017) is that they are "salvaging" these traditions from being lost, and by designating and presenting them in a mappable way, current and future people might be able to enjoy and learn from them. Scholars of performance and heritage have been critical of this approach. Performances of intangible heritage which tourists come to witness and take part in are often exploited by profit-seeking elites and used to obscure painful implications of colonial occupation.

One such case of performance for the global tourism heritage industry is in Cambodia, where the Royal Khmer Ballet performs in Cambodia's most iconic heritage landscape, Angkor Wat (Figure 5.2). Michael Falser (2014) explores how these performances impact national narratives in Cambodia, by threading the French colonial assessment of Cambodian tradition, through the post-colonial nation-building era, to the current Khmer Ballet performances which enact a pre-colonial narrative of Cambodian national identity for a global tourist audience.

Nationalism and Sport

Sport is a highly speculative performance, utilizing the body, crowds, landscapes, and ephemera to make statements, many of which are political. While sport can be engaged with at many scales, from the lone runner to the neighborhood basketball court to state and international championships and beyond, all can be tied to the nation and national identity. Many scholars of sport geographies have commented on the fundamental role that nationalism plays in the creation of global sport (Yan and Watanabe 2014; Bale, 2002; Tervo 2004). Likewise, the performances surrounding sport continue to reaffirm the nation's role through the actions of those who participate through watching, playing, or otherwise engaging with sport (Wise and Hall 2017). It also reaffirms the role of scale in national identities.

Much sports participation occurs on the local scale in neighborhood or school leagues, but most sporting attention, in terms of viewership and economic generation, occurs at regional, national, and international scales. Global mega-sport events are economic tent

Figure 5.2 Royal Khmer Ballet
Source: Daniel Mennerich

poles for media outlets, franchises, and athletes themselves, but are not necessarily profitable for the cities and countries that host them, or indeed for the masses that participate by watching. So why have these massive sporting events proliferated over the 20th century, and why do they have such enduring popularity? Other than war, no other competition has the potential to unite national groups as strongly as sport (Bale and Sang 2013).

Case in point is the rivalry between Real Madrid and Barcelona FC which developed over the course of the Franco dictatorship in Spain. These two football clubs have represented two sides of a contested nationalist story. Twentieth-century Spain was characterized by civil war and an ensuing totalitarian regime led by military leader Francisco Franco. Franco sought to unify Spanish national identity by stamping out regional cultures, including languages, religious traditions, and any regional aspirations for autonomy. Real Madrid, the most successful club in the world, was the flag bearer of the Spanish nation, Franco's team. FC Barcelona represents the struggle for autonomy within Catalonia.

Since the death of Franco this rivalry has grown with Catalonian protests in the 21st century. The meeting of these two teams is a platform and amplification of national tensions between Bourbon and Francoist Spain and independent-minded Catalonia. The 2012 rivalry game provides the clearest example, as described by Sid Lowe in his 2014 book *Fear and Loathing in La Liga*. The game took place in Barcelona, in the midst of economic crisis, failing fiscal negotiations between Madrid and Catalonia, daily protests and growing calls for an independence referendum. Barcelona fans covered the stadium seats with hand-held red and yellow cards, and the Catalonian independence flag with a yellow star on it was seen from most angles. Many expected the Catalonian "national" anthem to be sung, although it was not. At 17 minutes and 14 seconds into the game, a time stamp symbolizing the 1714 fall of the city to Bourbon rule, the stadium erupted into chants of "*Indipendencia! Indipendencia!*" This moment was described as prophecy – that Catalonia would have independence when Camp Nou, FC Barcelona's iconic stadium, demanded it (Lowe 2014). The years since that game have seen more Catalonian protests and a suppressed referendum on independence. Such is the power of sport to create and change political narratives.

The most noteworthy of these global mega sporting events is the modern Olympic Games. Resurrected in 1896, the modern Olympics were designed to comment on national (and by extension racial, cultural, and political) superiority (Berg 2008). Baron Pierre de Coubertin, who established the modern games, wrote about a forum in which internationalism could thrive, where young (male) athletes could meet members of other nations in competition (Bairner 2008). Yet the modern Olympics were also an antidote to a changing geopolitical reality. European colonies were being lost, and nations were realized as integral components of a global political order.

At this time there had been a generation gap in armed conflict, changes in the nature of economic production, and new administrative capacities extending the reach of the state (Winter 2003). These changes spurred what would come to be understood as a "crisis of masculinity" in the late 19th and spilling into the 20th century (Kimmel 2018). This can be tied directly to the establishment of collegiate sports in the Ivy League in the US, as well as the invention of American football. De Coubertin himself commonly wrote about concerns of French male fragility in the wake of wartime defeat, and just as Teddy Roosevelt was espousing the benefits of rough physical contact for young American men (Villanueva 2020). The focus on war and sport played an influential role in shaping masculine sensibilities of national belonging, and sidelining those who were not featured in these performances (Hogan 2003).

The Olympics have been seen as a "pure" national competition where groups of national athletes gather to represent the strength of their national communities (Brohm 1978). Along with focusing on the quality of individual performances, international sporting events such as the Olympic games also reinforce national boundaries (Bryson, 1987; Hargreaves, 1985; Rowe 2004). The sporting imagery that the Olympic games produce every two years reproduces the segmentation of the world into discrete national groups (Hoberman 2020). Sporting events such as the Olympics are one of the few legitimate and celebrated ways that nations promote their identities and agendas to a global stage.

Sport and sports journalism surrounding the Olympics helped establish a Finnish national identity. In the early 20th century, Finland was emerging on the global stage as an independent state, and thus as a new national unit. The Olympics proved to be an opportunity to distinguish Finland from neighbors (and former colonizers) like Sweden and Russia (Tervo 2002). The international sports movement was new at the time, and this allowed Finland to utilize mass media to compare itself to other nations.

Today, sporting excellence on the Olympic stage represents national excellence on a global stage, acting as a stand in for international strength and dominance where wartime conflict once did. Presentation and promotion of national excellence is especially highlighted for host nations, which in the past have been relegated to Western countries, perpetuating a core-periphery model (Huntington 1993). While Western countries continue to monopolize mega-sport coverage, emerging states have been selected as hosts more recently, and have been able to capitalize, if not economically, then geopolitically, on the sporting world's spotlight.

Narratives of the nation most clearly emerge during the opening and closing ceremonies, and during medal awarding ceremonies. Such ceremonies are required for Olympic participation and constitute a highly commercialized expression of national origins and identities. While each national team is present, the ceremonies focus on the national story of the host country. The choices made by the host country during the ceremonies might reflect political ideologies, real or imagined histories, or they might be aimed at promoting tourism or international cooperation and development (Hogan 2003).

China was selected as the host country for the 2008 Olympic summer games, following its East Asian neighbors Japan and South Korea. This was a watershed moment in China's global development and fulfilled its quest for modernization (Xu 2006). An analysis of the Beijing opening ceremonies shows the unified and Sinocentric Chinese national identity presented to the world on this special occasion (Lee and Yoon 2017). That opening ceremony depicted scene after scene of harmonized and unified masses of Chinese performers engaging in both pre-modern and modern arts, crafts, and pastimes (Figure 5.3), linking China's many historical achievements with its modern economic success (Lee and Yoon 2017). Themes included calligraphy and writing, militaristic drumming, and traditional dance. One through line in this depiction was a focus on historical Chinese ethnic nationalism.

Conversely, London's depiction of national identity for its 2012 opening ceremony focused on civic nationalism and (to an extent) inclusivity within a diverse and individualistic United Kingdom (Lee and Yoon 2017). Rather than an overwhelming mass of harmonized dancers, workers, or drummers, London flooded its ceremony stadium with slices of culture taken from various spatial and temporal corners of the UK, reflecting the modern diversity of stories produced there. Themes included literature, pastoralism, the industrial revolution, British comedy, and the National Health Service (Figure 5.4).

Figure 5.3 Beijing Opening Ceremony
Source: US Army, CC BY 2.0, https://creativecommons.org/licenses/by/2.0, via Wikimedia Commons

Figure 5.4 London Opening Ceremony
Source: Sum_of_Marc

Together these scenes portrayed a heterogeneous and "cool" British national identity. Both the 2008 Beijing and 2012 London opening ceremonies showed in separate ways how different national identities are constructed through key depictions of a national character, history, or culture. Global sporting culture which is epitomized by the Olympics, World

Cup, and others are now accompanied by mega media events, which facilitate these narratives both within and outside of the location they are held.

The Olympics and other sporting events can provoke controversy in terms of who speaks for the country. Singapore won its first Olympic medal as an independent state in 2008, at the Beijing summer Olympics. However, the win was devalued because the athletes who had won the medal were Chinese born, and not accepted by Singaporeans, many of whom are ethnically Chinese, as national representatives. The athletes, Li Jiawei, Wang Yuegu, and Feng Tianwei, had become naturalized citizens in 2006, previously recruited by the Singaporean government in an effort to bolster Olympic success and subsequently national identity. Video of Beijing-born Li talking about playing on "home turf" and dragging the Singaporean flag on the ground during the opening ceremony angered many Singaporean observers, who lit social media ablaze questioning the national allegiance of these foreign players. While this could be viewed as no more than a public relations debacle, this story illustrates the importance of sport in nation-building efforts, the sacredness with which the sporting arena is treated, and how easily sport is used as an example of national character. This story also shows the importance of sport from both a top-down and bottom-up perspective. Singapore made the conscious decision to recruit foreign talent for their national team, betting that a win at the Olympics would bolster national morale. What happened, however, shows that perceived national authenticity is also important when it comes to sport.

Personifying the Nation

National identities are often personified in characters, such as the farmer (Little 2002), the cowboy, explorer, or pioneer (Rojas 2013), the sports hero (Parry 2021), or the troubadour or dancer (Mulholland 2007). Sometimes these characters are nameless folk personas, representing a well-known national trope. Sometimes they are allegorical characters, personifying a stereotypical or aspirational national identity. These characters are given life through stories in literature, poems, songs, and oral history, and recreated and replicated through visual mediums like illustrations on maps, comics, and paintings, and through performances in local theater, movies, and TV. They represent how nations can be read and understood by those both inside and outside of the nation. At the same time, they solidify gender norms, racial boundaries, and class stratification within a given nation, providing a well-ordered picture of national identity (McCallum 1997).

In many countries, sport is so intricately linked to national identity that a particular team, style of play, or individual athlete will commonly be upheld as the prime example of national excellence and uniqueness. In Argentina, the development of a unique style of football (soccer) in the early 20th century helped to establish a cultural bond and a national identity among a country of recent immigrants (Archetti 2017). The cult of Argentinian football grew out of the emergence of a few key aspects which set Argentinian players apart from European players. Players such as Diego Maradonna came to represent a stand in for Argentinians on a global stage when they won the World Cup in the 1970s and 1980s. Maradonna and more recently Leo Messi, the most highly paid footballer of all time, are the most famous Argentines of all time along with political leaders Juan and Eva Perón. The style of play that Maradonna and his teammates exhibited served as a marked difference between them and the rest of the world. The dribble and use of pairs was a style that is still defined as "elegant, skillful, cheeky and lively" (Archetti 2017), attributes that would go on to be applied to other aspects of Argentinian society.

Argentina at the turn of the century was a new country experiencing a massive cultural shift fueled by immigration and economic prosperity. It was also a country in search of a national identity. The Gaucho emerged as the emblem of Argentine nationalism, representing ruggedness, masculinity, and pride in a mythologized Argentine past (Archetti 2007). The nameless cowboy figure of the Gaucho emblazoned products and defined them as "authentically Argentinian," and Gaucho was used as a descriptor for any "typically Argentinian" style. Polo players and footballers were noted to have "Gaucho flair" in their moves, and elements of Gaucho clothing became part of the national costume (Figure 5.5). Paradoxically, the

Figure 5.5 Gaucho in the flag day parade in Rosario, Argentina
Source: Pablo D. Flores, CC BY-SA 2.5, https://creativecommons.org/licenses/by-sa/2.5, via Wikimedia Commons

predominance of the Gaucho character in Argentine nation-building came when actual cow wrangling in the rural Pampas was becoming an extinct profession.

In addition to clothing, the Gaucho became inextricably linked to the preeminent Argentine creation: the Tango. Developed in the outskirts of Buenos Aires in the 1880s, an abundance of *porteño* cafes, clubs, and cabarets ushered this modern style of dance into the heart of the city where it captured the Argentine imagination. Technologies such as the radio and television exported the Tango around the world in the 20th century, and the dance, music, and narrative of the Tango became the cultural products most strongly associated with Argentina. Perhaps as a result of Europe's external gaze (Archetti 2007) or as a backlash against modernization (Bockelman 2011), the Tango came to incorporate elements of the Gaucho, blending rurality into this fundamentally urban artform. Use of the bolero and wide brimmed hat were brought into some performances, and Tango composers and dancers incorporated more Gaucho stories and imagery into their performances. Both the Gaucho and the Tango are "performed" across the Argentine nation, highlighting both the reality of urban creativity and the rural idyll as personified through dance, song, caricature, and costume (Bockelman 2011).

Musical performance can signify a national characteristic, as mariachi unified a national imagery under the ideals of *machismo* and *mestizaje* in Mexico, the mixture of Indigenous and European ethnic and cultural traits (Mulholland 2007). Mariachi is an ensemble usually played by 8–12 men, dressed in *trajes de charro* (silver flecked cowboy costume), and adorned with *sombreros* and cowboy boots (Figure 5.6). Mariachi songs encompass a pantheon of foundational Mexican myths, and constantly reference and build upon themselves, lending an historical authority to the musical tradition as foundational to the essence of Mexicanness.

Mariachi was promoted by the Mexican government after independence, at a time when regional arts and culture as well as Indigenous pride were highlighted as elemental pieces of Mexican national culture. Although originally from Jalisco, mariachi and the men that performed it soon became synonymous with being Mexican in general because of its use as both a central performance and a musical backdrop for radio, film, and television. While mariachi

Figure 5.6 Mariachi band
Source: Matt Gush, CSUF Photos

has become ubiquitous at restaurants, resorts, and other "Yankee" tourist spaces, Mexicans also readily participate in and consume mariachi from small towns to big cities, where it is performed at weddings and quinceñieras, town celebrations, and other parties.

Male and Female Symbolism

Gender plays a key role in the way people experience and participate in nationalism, as well as in the establishment of each state or nation's own national identity. States have used gender roles, usually in unison with religious understandings, to solidify state identity and promote nationalism to exclude "others." Women have traditionally been used to gauge the morality and purity of a state, while men often portray its strength and honor. Nationalistic goals shape these gender roles and assign them in concrete ways to create a cohesive state image and power. Nationalism itself was an ideology influenced by an international system built by and for men; as such, male gender roles closely reflect this ideology while female roles have traditionally been in support of men working to uphold the nation. This includes anything necessary to support the nation, often as reproducers and teachers of nationalist norms who work domestically to ensure that the nation can continue (Mosse 1985; Nagel 1998; Nagel 2008).

Many of the male characters draw upon a masculinist nationalist discourse like the cowboy, the pioneer, and the warrior. These symbols show men constantly competing to show themselves as worthy, powerful, and honorable. Men are often expected to participate in high-stress military and government positions as fighters and ultimate decision-makers who proudly push the nation forward (Nagel 1998). Likewise, female characters in national mythology often embody hyper-feminine qualities, and are often portrayed as caretakers, mothers, or virgins. These female national characters are both strong and subdued, appearing more often in paintings, statues, and as allegorical figures than as physical performance figures. Among the best known of these is Marianne, an icon which replaced the monarch during the French Revolution of 1789 (Figure 5.7). Marianne represents the French ideals of liberty, motherhood, and citizenship (Mitra and König 2013). In a painting by artist Eugene Delacroix, Marianne is depicted as the standard bearer of the French nation. As Mitra and König (2013) point out, Marianne has been used to serve all kinds of purposes, from the embodiment of a traditional France – as employed by the ethnonationalist National Front – to a continued symbol of republican values.

Both male and female sexuality is used in efforts to vilify the enemy, further defining national gender roles as a response to others. Female sexuality is controlled by a system of patriarchy that holds women up as being bound to their male counterparts, instead of having the ability to make their own sexual choices. "Proper" female sexuality is bolstered within the national identity, depicting obedient, heterosexual women, while attempts to weaken others present women as disobedient, immoral, sexually loose, and manipulative (Nagel 2008). While women are often the victims of greater controls, gay men are also at risk of damaging national image, forcing them to live under great scrutiny and fear of persecution. The fear of LGBTQ+ rights is most present in Eastern European countries, where LGBT participation is low and anti-LGBT legislation curtails advocacy. Russia and Hungary are both examples of countries that view LGBTQ+ advocacy as a threat to the national model and have taken steps to criminalize public displays of non-traditional sexual orientations (Nuñez-Mietz 2019).

Figure 5.7 Statue of Marianne in Place de la République, Paris
Source: Dave Kaplan

Territories were often personified as helpless or naive young women during colonization, providing a visual for the idea of "virgin land" and a justification for the imposition of a patriarchal colonial structuring (Warner 2000). Fraunhar (2005) explores this trope in the Cuban mulatta, who is situated as a national symbol and a sexual being throughout the 19th and 20th centuries, who, unlike the Indian or Black woman, is a hybrid persona who is "like but not quite" (Bhabha 1994) the colonial male who gazes upon her. Fraunhar (2005) shows how the mulatta comes to symbolize not just Cuba but the whole Caribbean through cinema.

The symbolism of women as mothers was used in Ireland as it sought independence from England (Cullingford 1990). Allegorical "Ireland" changed at the turn of the 20th century from a young woman to an old lady, imbued with domestic knowledge and a strong connection to her figurative children. In *Cathleen ni Houlihan*, a 1902 Yeats play, the maternal figure came from Celtic mythology, centering Celtic history within Irish nationalism. It suggested that the nation could now protect itself, solidifying a new sense of national identity among Ireland's masses (Kitishat 2019). The symbol of motherhood continues to be a fraught one in Ireland as sex and pregnancy became outright taboos during the Irish nationalist period, and a new female symbology, one of the "Irish Madonna," emerged (Morisson 2019). Generations of Irish women have become more open about countering this national narrative through art, protest, and legal challenges. In 2018, abortion became legalized in Ireland, a recent step in the detangling of motherhood and Irishness.

The Role of Rurality

Performance of folklore is fundamental to the idea of performative place and nation making (McKay 2009). Because folk culture is so tied up with national culture, the branding associated with rurality is overlaid on a national brand. Rurality, including the people that live in rural places, rural landscape, and jobs tied to rural life are glorified as national ideals. People and places outside of elite cosmopolitan influence are purported to make up the "real" stuff of the nation.

New Zealand, for example, frames rurality as the "backbone of the nation," idealizing the work ethic and independent entrepreneurialism associated with pastoral work (Bell 1997). Images that perpetuate idealized visions of rural New Zealand life abound in popular culture, from murals to postage stamps, to television shows, to festivals (Bell 1997; Fountaine 2020). Reproducing the rural idyll in New Zealand is mediated through these channels, which people consume as both factual and mythologized.

Fountaine and Bulmer (2022) explore these themes by analyzing the television show *Country Calendar*. This show depicts rural life and promotes farmworker voices with real implications for agricultural policy in New Zealand. Most of New Zealand's population live in urban centers, yet agriculture makes up a sizable percentage of the economy. *Country Calendar* mediates what has been portrayed as an increasing urban-rural divide in New Zealand, reminding urban dwellers of the importance of agriculture in a very deregulated market. Likewise, computer games that recreate farming life have proliferated in New Zealand, simulating an idyllic rural landscape for urbanites to enjoy (Sutherland 2020). Stories and images like this shape public opinion of rurality, framing a rural lifestyle as aspirational and nostalgic, subtly reproducing the nation by recentering foundational myths of transforming and taming nature and rugged individualism.

National Performance through Policy

Nation-building and policy go hand in hand. Policy, whether official or unofficial, has a territorial delineation, applicable in certain areas, namely within "national" boundaries. Civic conceptions of national identity use policy to promote national togetherness; in such cases, members of the nation are seen as anyone who properly performs the behavior dictated by national policy. However, we can also point to ethnic nationalism in many of the performances that are generated through policy, like speaking a certain language in schools, or placing a hand on the bible in a court of law. These political traditions are conscious efforts on the part of elites to create a culture which supports the political system that they govern

(Gellner 2003). Citizens perform national policy through such mundane acts as standing for and reciting the Pledge of Allegiance, which links everyone in the nation to the ideas embodied in the constitution, yet certain ethnic traditions are privileged through policy, even in a multi-ethnic state, when some religious holidays are federalized and others not. Fundamentally, elites make political decisions to generate a sense of togetherness, drawing from a variety of histories, ethnic and civic traditions, and tying this group identity to a territory. The territorialization of tradition is fundamental to the essence of nation-building.

People might become citizens through the practices of traversing everyday landscapes in ways the state has deemed suitable (Scott 2008). This territorialization is fundamental to national identity and lays out all of the formal and informal rules of interacting in and with a given place (Sack 1986). These everyday practices are national performances too. Stopping for a red stop sign. Queuing up at the grocery store. Speaking Norwegian instead of Sami. The policies that are deployed through state territorial apparatuses set the stage for the performances of nationalism that follow (Butler, 1997; Sioh 2010).

Joseph (1999) describes these policy-set stages as "Citizenscapes" where people embody participatory politics based on the socially and politically constructed ideologies of belonging to a nation. When viewed in this way, everyday actions like walking on the left side of the sidewalk instead of the right can be seen as performances conditioned by state policies.

The act of driving a car is central to both British and American national identities, and has been conditioned by a matrix of material, political, and cultural national identities (Edensor 2004). Various kinds of cars, driving styles, sizes, and decals are given different personality types in many cultures. Driving emerged as a pastime of British elites in the interwar period, and intensified through popular portrayals of intelligent, sophisticated British men in smart and sleek cars. In the United States, autocentricity stems from national values like rugged individualism and free market capitalism. Other aspects of mobility, such as mass transit and pedestrianism, are often marginalized as most plan their days around cars. Moreover, automotive brands are associated with certain groups through popular culture and stereotyping. These range from a home-built derby cars and lifted pickup trucks for young rural men (Thompson and Tian 2008), to urban lesbians and their Subarus (Sender 2003), to import streetcar racing among Asian Americans (Kwon 2004). A car purchase denotes where one fits within the national fabric.

Some post-colonial states deploy policies which encourage performance of a national identity which might be read and recognized by those Western countries in power. Post-colonial states must affirm their right to exist in a geopolitical sense, and this is done through multiple, repeated, and compulsory performances of national belonging. These policies constitute strategies which range from promotional to punitive.

In post-colonial Malaysia, transitional anxiety contributes to nation-building through the creation of institutions and policies which create spaces for performing national identities (Sioh 2010). Such performances were enforced within panopticonic "new villages," where daily routines were monitored. By creating new landscapes through which citizens might be more visible, the Malaysian counterinsurgency government enforced performances of a post-colonial identity by forcibly displacing people, enacting curfews, and promoting agricultural schemes. Within this curated landscape, a new Malaysian national identity was formed and essentialized under the strict guidance of the state, replicating itself through the daily actions and interactions of citizen-subjects (Sioh 2010).

Box 5.1 Nationalism in the Military

Figure 5B.1 Nationalism in the military
Source: Library of Congress Collection, US WWI Posters

If nationalism is defined as the connection to an abstract idea of the nation's existence, patriotism is a direct defense of the state when it is threatened (Snyder 1954). As a source of power for the nation/state, the military is an embodied representation of both nationalist and patriotic efforts. The size and power of the military is often used to measure a state's own potency. Military actions are justified when the life of the nation is threatened, and great honor is conferred on those who participate in these actions (Doob 1976; Nagel 1998).

The military also serves as a nationalizing institution (Snyder 1954). In a volunteer military, troops will possess necessary levels of nationalism and patriotism, heightened by the military's own rituals. A conscript military begins with less motivated members but can bring together disparate people in service to a common goal. The invocation of nationalism, especially during times of real or perceived peril, promotes military participation and assigns honor to those who answer the call.

A strong military force works from an idealized image of what the nation is and represents, compelling service members to pledge undying loyalty to the nation-state no matter the costs. Fighting for one's nation is seen as a very honorable way to live, and the lack of willingness to fight becomes selfish (Snyder, 1954). The military also imbues its forces with images of what the nation is not, allowing antipathy to become an important emotion in establishing and maintaining nationalist efforts (Snyder 1954). Propaganda paints a negative image of the "other," including questions of faith, sexuality, and morality (Nagel 2008). Especially during times of war, nationalist sentiment becomes easier to spread as the population's fear of losing the nation increases.

Competing National Performances

Mainstream performances like fourth of July parades in the United States and Bastille Day celebrations in France rarely deviate from long solidified and universally understood versions of national performance and symbology. Yet there are other celebrations which present contrasting versions of nationhood which can be produced via the visible and highly political process of public performance. Public rituals heighten boundaries between minority and majority populations, social classes, as well as linguistically, religiously, and ethnically diverse groups. Performances of alternate versions of nationhood can coexist peacefully, or they can rival each other in overtly aggressive ways. Cohen (2020) describes these alternate articulations of national identity as struggles of "ownership" over the symbols of the nation. Visualization of the social structures at work manifest fractures in the group identity of the nation, causing crisis within the mainstream.

White nationalism in the American South has long been normalized through performance at almost sacred historical spaces, like plantations (Bonds and Inwood 2016). White Americans make pilgrimages to plantations such as Natchez in Mississippi, where the hierarchy of white elites and invisible subjugated people of color is acted out for audiences (Hoelscher 2003). The racial hierarchy of white American nationalism is reproduced through performances like these, ingraining a sense of historical superiority among the white majority and occluding the painful experience of Black people in the United States and the legacy of slavery in modern day America (Alderman 2010). Thousands of tourists journey to plantations such as Oak Alley in Louisiana to immerse themselves in the perceived romance of the (white) antebellum south (Figure 5.8). An increasingly educated and socially conscious population has started questioning this dominant historical narrative to understand how the plantation, and the American South in general, encompasses several contested national narratives (Alderman 2016; Alderman and Modlin 2014; Cook 2016).

Figure 5.8 Oak Alley Plantation, Louisiana
Source: Michael McCarthy

In Cambodia, memory of the 1970 genocide is still largely contained within two official memorial sites, even though the genocide by definition was not a spatially contained event. Tyner et al. (2012) explore how unmarked and unremarked upon spaces of violence affect memory of the genocide among locals and tourists in Cambodia. Engagement with the genocide period is part of a growing dark tourism agenda in Cambodia (Tyner 2020). This dark tourism agenda includes visiting Tuol Sleng Museum of Genocide Crimes and the killing fields at Choeung Ek. The isolation of these sites from the capital city means that many Cambodians might not be able to engage with these histories, and tourists will have a skewed interpretation of the ubiquity of the events. Knowing the spatiality of the genocide is key to understanding how it has affected national identity in Cambodia (Tyner 2016). Engaging with memorial landscapes is a form of public pedagogy, especially in a country with a large rural populace (Hannum and Rhodes 2018). Gomez (2021) shows another method that local individuals might take some agency in remembering the Cambodian genocide, by mapping grave and prison sites across the country.

The performances of national identity can sometimes conflict not just based on social belief but on local context. Among Northern Irish Protestant marching bands in Northern Ireland, British identity is displayed in an unambiguous and literally out loud fashion (Casserly 2018). However, depending on the local context in which the band performance is being held, the identity conveyed in the performance itself can shift in multiple and often conflicting ways. This is true in many examples. When in unsafe or unsure spaces, individuals and groups performing national identity must consider the potential consequences of conflict.

While individuals within national groups are logically considered active agents, able to make their own decisions and form their own beliefs, historical memory and narrative undoubtedly inform understandings of current geopolitical relationships. National belonging and othering might hinge on storytelling from a certain point of view, leading generations of young people to carry certain beliefs (Popov and Deák 2015).

All national identities contain some contested stories about historical events, leading to certain beliefs about their national group. However, some people grow up enveloped in such contestation. In the Falkland Islands, youth are presented with narratives and counternarratives of key national events daily, namely the attempted liberation, or occupation, of the Falklands, or *Islas Malvinas*. Argentina's continued claim on the islands informs popular conceptions of Argentine national identity, while the islands are legally British Overseas Territories (Benwell 2016, 2017). Recent developments in this territorial contest include the discovery of oil in the surrounding waters, a resurgence of media attention surrounding the 30th anniversary of the 1984 war, as well as a governance referendum designed to put the question of jurisdiction to rest (Dodds and Hemmings 2014). Argentina's tireless efforts to reclaim the islands present contested notions of national belonging for islanders, who are caught in a symbolic geopolitical struggle. While symbolic on the international scale, the contested nature of the islands presents challenges for those who live there, as they are forced to navigate these diametrically opposed claims on nationhood, as well as be the subjects of militaristic posturing across two popular and political cultures (Benwell 2014; Benwell and Pinkerton 2020).

Agency and Language

Speaking or writing a language is the ultimate everyday performance. It is how people communicate and understand the world and one another. Language stands apart from other cultural markers due to its necessity for group organization and its adoption by institutions such as governments. Language is also uniquely positioned to betray an individual's group belonging. Try as one might, accent, dialect and fluency immediately let others know if you are a part of or separate from the local group.

Language is also a particularly "hot" performative expression of national identity because there are thousands of languages and only around 200 states. So, the vast majority of languages are being spoken in a minority, unofficial, or a sub-state context. Because language is a key marker of ethnicity, it becomes central to ethnonational campaigns for autonomy, recognition, or separatism. The political success of nationalist movements often hinges on language mobilization. And conflicts come with states who have a vested interest in elevating one language at the expense of all others.

Spain is a case of a state with multiple sub-state nationalities that hinge on ethnic belonging. Linguistic performance is paramount to belonging in these nations. In Catalonia, use of Catalan has been central to the Catalonian push for independence for over 100 years (Conversi 1990). Language has been found to be the most important predictor of nationalist sentiment in Catalonia, even among those whose first language is not Castilian (Spanish) (Argelaguet 2021). This case is unusual however in that the regional language of Catalan has enjoyed strong social capital for generations, and there is a longstanding association between speaking Catalan in Catalonia and upward mobility (Della Porta and Portos 2021). This bodes well for the adoption of Catalan among newcomers, and for the strong mobilization of the language, leading to political success.

Other national languages in Spain, like Galician and Euskera (Basque), have taken different paths. Galician for example had been deemed a language of the poor, predominantly spoken in rural areas. Therefore, it suffered lower social capital and was associated with low social mobility (Beswick 2007; Máiz 2003). The Galician nationalist movement has not been as widespread or demanding as the one in Catalonia in the same time period. However, new speakers of Galician, or *neofalantes*, are primarily educated and young, showing how choosing a language at any given moment signals a political and national agency (O'Rourke and Ramallo 2015).

In Basque country, ethnicity has been the key point of contrast in the nationalist struggle, whereas language has only recently been centralized. This was due in part to lower speakership among Basques at the time of nationalist organization, as well as low language loyalty among Basque elites, and a multitude of variation among those who did speak the language (Conversi 1990). The ethnic focus led to a dissolution of the Basque nationalist movement into an armed struggle which captured negative international and domestic attention. In the 21st century, however, Basque nationalism, while still ethnic in nature, focuses more on language and the willingness of individuals to acknowledge and participate in Basqueness.

While language use as an expression of hot nationalism often occurs at the sub-state scale, where minority languages are vying for recognition and concessions in the rigid state system (Hannum 2020), language use in the creation of new supra-state nationalisms can also be found. An example of such an entity is the Arab national movement (Kramer 1993). Spanning from the Atlantic shores of Morocco to the Pakistan border, the unique states that compose the "Arab homeland" began to organize for international solidarity in the 1950s, in a push for a unified expression of nationalism among (most) Islamic and Arabic-speaking states. This was a response to 20th-century conflicts from European and Israeli interests that were seen as threatening not just one Arab state, but Arab identity itself. Rather than a supra-state entity created out of an economic or military alliance, this demonstrated the re-conceptualization of a nation where superimposed boundaries had been imposed. The Arab nation utilized a common religion, history, and language as basis for existence (Suleiman 2019). Language in particular was mobilized. However, this new Arab national identity did not arise without tension from individual state powers. Although the European-imposed boundaries were superficial, the new post-war Arab states that had been carved out of the Ottoman Empire developed identities of their own in addition to a larger Arab identity (Ajami 1978).

Conclusion

Nations are cultural entities. They can be promoted by the state and various institutions within the state. Nations can find expression in their territory and in their landscapes. However, it is through the daily practices of normal people that national identity is continuously affirmed. When describing nationalism as a "daily plebiscite," Ernest Renan was implying the sequence of performances put on by each member of the nation, showing how much they belong.

In this chapter, we have shown how nations rely to a great extent on a past, often imagined or exaggerated, but which hearkens back to ties that descend into the mists of time. Folkloric traditions are not simple cultural exercises. They are also affirmations of the ties between peoples in a nation. Parades, festivals, historical reenactments, music all contribute to these ties, creating a sense of common heritage and common aspirations. The sense of nation may be wrapped into a single iconic figure, such as the Argentine Gaucho or the French Marianne, which again portrays a set of values a nation claims to hold dear.

National performances are not only intended for internal consumption. They also become ways in which the state can project itself externally. One means is through tourism, in which visitors come to capture a national essence in just a matter of days (or sometimes hours) by experiencing a carefully curated landscape. Another is through international sporting events, especially the Olympics, where athletic prowess and national pride are conjoined. Plus, the host country gets a chance to use the Olympics as a stage to project their preferred national narrative.

While the national elites and the governments they run have an outsized role in shaping the national narrative, counter examples are found everywhere – sometimes secretly but also overtly in more liberated spaces. These may come from subnational movements or from subversive ways of interpreting national history. Such examples demonstrate that national identity is rarely simple and uncontested.

Chapter 6

Multinational, Federalist, and Supranationalist States

For most of history, all substantial political entities were multicultural. Ancient and medieval empires cobbled together dozens, hundreds, of distinct cultural communities under a single ruler. Smaller states, not grand enough to be considered empires, contained fewer distinct cultures but were also quite diverse. In the age of kingships, it was common – even expected – for the ruler to speak a different language than the ruled. The smallest entities, the city-states, regularly hosted different peoples they considered "nations" who inhabited distinct quarters of the city, analogous to today's patterns of ethnic segregation.

The organization of cultural diversity varied from brutal exploitation, to tribute, to some measures of participation. Most such polities recognized those cultural groups who occupied the top, others who were treated badly, and many in the middle. But diversity itself was not considered a political problem. This changed with the ideology of nationalism. Nationalism demands that polities line up with culture, leaving scant space for difference. It is in this vein that the earlier nationalist movements either sought to scrub away internal differences through education, persecution, and propaganda or they would seek independence from a polity with which they did not share any commonality. The rights of self-determination, pronounced most forcefully in Woodrow Wilson's 14 points, were a watershed in breaking up the various land-based empires preceding World War I (Ottoman, Austro-Hungarian, Russian) and providing nascent nations within these empires a doctrine for them to negotiate their own country.

A true nation-state is a chimera – something hoped for but nearly impossible to achieve. This applies especially to countries with substantive area and population. A few smaller island states – with Iceland the largest example – represent a rare breed of true nation-state. The vast majority of countries in the world must contend with significant cultural differences and develop tools of compromise or coercion.

The purpose of this chapter is to explore how cultural distinctions have or have not been accommodated. We examine the political models that have worked more or less effectively. Additionally, we discuss supranationalism, a phenomenon whereby independent states decide to band together and create an entity larger than themselves.

The Nature of the Multinational State

The idea of a multinational state is commonly discussed among political scientists and political geographers and refers to a country which contains more than one self-identified nation. As such, multinationalism is a type of multiculturalism. Yet whereas

DOI: 10.4324/9780429435782-6

multiculturalism encompasses a broad spectrum of cases at varying spatial scales, multinationalism is more limited to self-defined nations within a state.

The basis of distinguishing multicultural from multinational is that each nation within a multinational state considers itself to be culturally and politically distinct. Connor (1978) differentiates nations from ethnic groups. An ethnic group refers to a community which shares a common culture and heritage; a nation includes that common culture but adds the political goal that they deserve some degree of autonomy (perhaps even independence) and the opportunity to govern their own affairs.

Almost all countries of the world contain multiple ethnic groups, whether by virtue of established communities, of new immigrant populations and their descendants, or of old regional-national divisions that still have salience. Taking an example from the United Kingdom, British cities and towns include multiple ethnic groups: many from former British colonies in South Asia, the Caribbean, West or East Africa; from populations coming from other European countries; still others representing movements of people within the British Isles, notably from Ireland (Office of National Statistics 2003). Many of the categories derive from the British census. They allow for self-definition and categories are often combined to preserve confidentiality and reliability. Ethnic groups within Britain, as in most countries, have political goals. The nature of electoral politics in many countries harnessed the collective power of these diverse groups and most are eager for cultural recognition and certain civil rights.

The same British census also uses identities based on the UK's traditional nations of England, Scotland, Wales, and Ireland, though not on earlier distinctions like Cornwall. This recognizes the significant qualitative difference between nation and ethnicity in the British context. This distinction is also rooted in political aspirations. A nation like Scotland carries a unique political history and a future that differs from that of England (and by extension that of the UK as a whole which is nationally dominated by England). So even while members of the Scottish nation can disagree on the ultimate desired political outcome, most would agree that some form of political/cultural recognition for their nation is warranted. On the basis of this solidarity, Scotland has been able to extract several concessions from the British government (Keating 2012). It has a thriving national party, the Scottish National Party, representing most of Scotland's parliamentary seats in the House of Commons. A group that defines itself as a national group expands on its political aspirations. If allowed, this could include plebiscites on whether the nation should become independent. Scotland conducted such a plebiscite in 2014, in which the vote against independence prevailed 55%–45% (BBC News 2014). Other actions may be less peaceful, up to and including civil war.

The geographical limitation is that each nation within a multinational state defines itself territorially. This could potentially be disputed. For instance, diaspora populations may consider themselves nations but here the territory is outside the host country. Most of the time, a nation predominates within a region and feels that this region, with some modifications, will serve well as the basis for an autonomous or independent state.

The messy cultural makeup of the territory can presage any turmoil if a breakup occurs. The fragmentation of Yugoslavia proceeded differently within each separate republic. Much of this had to do with the mixes of population. Pre-fragmentation, the Republic of Slovenia contained a population that was primarily ethnic Slovenian. Slovenia was also the first region to declare independence. There were a few skirmishes but the succession went fairly smoothly (Gow and Carmichael 2000). Contrast this experience to that of the Republic of Bosnia-Herzegovina, with its mixed populations of Bosnians, Serbs, and Croats. The desire of Bosnians to form an independent state collided with the goals of Serbs who also lived within Bosnia, many times sharing space and community life with neighboring Bosnians. The fallout from these differing aspirations gave rise to a years-long brutal conflict which has yet to be fully resolved (Oluic 2007).

Types of Multinational States

While we can easily count the total number of sovereign states, we can only provide vague estimates on the number of nations, depending on which criteria are used and how strictly these are applied. Even the lowest range of the number of nations (about 2,000) is ten times greater than the number of states (about 200). So, the default position of any country is to be multinational. But multinational states take many forms (see Figure 6.1). In the first instance are countries thoroughly dominated by a single national group. The state territory is made up primarily of a dominant nation, with only tiny pockets where separate national groups reside. A second example would be a country where a single national group is clearly in the majority, but one or more prominent minority nations occupy substantive spaces and contain significantly large populations. The final example occurs where a multinational state contains no clear majority. It could be divided between two or three large groups, it might comprise dozens of national groups, or there can be any number of permutations in between. In these examples, the fragmentation of national populations does not usually mean fragmentation of power. While these multinational states may have no national *demographic* majority, members of a single national group could still effectively dominate the country.

There are many examples of the first type of multination, where minority national groups are few and quite marginal. Japan is often held up as an example of a nation-state, and in truth the Japanese national group dominates this island-based country. The development of a strong Japanese national consciousness developed over an extended period of time and has consolidated into a strong belief in their own sociocultural uniformity. Japan's loss in World War II may have shifted the position of the emperor to something more ceremonial, but the emperor remains as a symbol of the Japanese nation (Yamazaki 2008). Yet this also glosses some differences. Japan has a fairly low proportion of immigrants, yet ethnically distinct Koreans and Chinese are present within Japan, many still not eligible for citizenship after two or three generations (Yamazaki 2008). The Japanese burakumin remain as a distinct outcaste group (Neary 2008). However, these

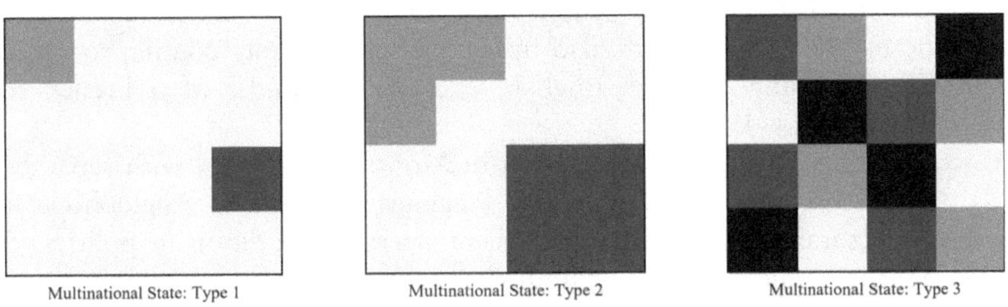

Figure 6.1 Multinational types graphic
Source: Michaela Gawrys

groups do not represent nations. Strong regional/national identities, separate from the Japanese nation, can be found at both the northern and southern extremes of the Japanese island archipelago. In the south, a set of islands that long was part of the Ryukyu Kingdom developed a distinct national identity before being incorporated into Japan in 1879. This region, with Okinawa as its capital and largest island, was again separated from Japan and occupied by the United States from 1945 to 1972. While several residents of these islands share in a larger Japanese identity, there is also a distinct ethnic identity based on language, appearance, and custom, and a long heritage of independence (Yamazaki 2008). On the northern island of Hokkaido, the Ainu were once considered an inferior, racialized population (Siddle 1996). Any claims they held onto territory were dismissed. Their land and resources were quickly taken during Japan's Meiji era. Yet the Ainu have come to reassert themselves as a distinct people, in league with other indigenous groups. Neither the Ryukyuan or Ainu represent a serious challenge to the Japanese nation, yet they expect some level of acknowledgment and understanding.

Many countries in the world include a dominant nation but one or more strong minority nations, each of which occupy large chunks of the state's territory. For the purpose of a state's internal security and cohesion, these minority nations must be managed differently than more marginal nations. Examples around the world abound, and we might distinguish between those states with a large majority nation and just one or two large minority nations and those states with a multiplicity of different nations but a majority or plurality nation which is clearly larger in size than the others.

Canada is an example of a state that includes a majority nation, a potent minority nation, and some smaller national identities among indigenous groups. Since its founding as a British colony and then an independent country, Canada has been unequally divided between the English-Canadian majority and the French-Canadian minority, with most Francophones (speakers of French) found in the Province of Quebec. During the 19th and into much of the 20th century, Anglophones controlled society and the economy in most spheres, including in Quebec's main city of Montreal.

Montreal represents a classic example where two nations developed separate systems while cohabiting the same city. Long the economic center of Canada and the largest city in Quebec, Montreal's economy was dominated by Anglophones (as English speakers are called). The major companies based in Montreal, including Seagram's and Molson, large banks like the Bank of Montreal, and universities such as McGill operated in English (Levine 1991) despite that fact that Francophones constituted a majority of the city and a huge majority in Quebec itself. Among the grievances of Quebec's French-speaking population prior to the 1970s was the extent to which they would be forced to communicate in English within the bigger cities and in order to obtain more desirable jobs (McRoberts 1988). They were also upset over immigrants coming to Quebec but overwhelmingly choosing to learn English, even in the midst of a French majority population (Castonguay 1987).

There were strong patterns of segregation in Montreal early on, with separate neighborhoods or suburbs predominating in one language or another. Anglophone interests developed an electoral system which gave them overrepresentation in politics (Sancton 1985). Separate media allowed (and still allows) each group the opportunity to get their news from dissimilar sources. A set of parallel educational establishments was likewise developed, absolving Anglophones from any need to learn French and, most troubling to the Francophone majority, capturing much of the large immigrant population.

That began to change in the 1960s and 1970s as Francophones asserted their national rights and formed strong nationalist parties. The result has been the devolution of many rights and responsibilities to Quebec and its Francophone government and continued movements for independence, none of which have succeeded. Yet the notion that the Francophone population constitutes a nation is well understood by all (Kaplan 1989). Figure 6.2 is a photo posted by Canadian Prime Minister Justin Trudeau meeting with the leader of the Quebec Avenir Coalition in 2013. Trudeau posted the image with a bilingual caption, indicative of public concessions made by the English-speaking majority in Canada.

There are also many indigenous groups within Canada, considered as "First Nations" and provided with the appropriate acknowledgments. Of these, several reservations had been established but the most notable development has been the separation of the territory of Nunavut from the Northwest Territories in 1999. This is the largest territorial subdivision in North America (not including Greenland) but has a tiny, Inuit population. Unlike in Quebec, there is no serious independence movement; further autonomous control over the Nunavut territory seems to suffice for the moment.

Indonesia furnishes a case of a state with a dominant nation (Java) and several minority nations. More than many multinational countries, Indonesia has developed a strong state identity, based around its common experience as a Dutch colony as well as a set of guiding principles. This is impressive given the absence of any real past political unity. It is "a twentieth century invention" (Webster 2007, 86) that has developed a strong civic nationalism but with the ethnic Javanese at the center. The heartland and dominant region of the country is the island of Java, which contains Indonesia's largest city (Jakarta) and more than half the total population. Beyond this are some 15,000 islands

Figure 6.2 Justin Trudeau meeting with Quebec Avenir Coalition Leader Legault. The Quebec Avenir Coalition is the leading provincial nationalist party
Source: Justin Trudeau

which contain clearly defined national groups. This provides not just remoteness from the state core but the potential for a powerful regional sensibility bound up in the island nature of minority national territory. While one national region of Indonesia, Timor-Leste, did become independent in 2002 after years of conflict, other strong regional/national identities – such as Aceh, Bali, and Papua – have remained as part of Indonesia (Wood 2008). The powerful sense of difference remains, however. Aceh, the westernmost part of Indonesia, is strongly Muslim and has a previous heritage as a sultanate. It has long resisted Javanese encroachment and spawned resistance movements against the "foreign regime" of Jakarta and the "alien people" of Java (Webster 2007, 90). On the other side of Indonesia, on western New Guinea, a sense of Papuan national identity that spans a multitude of tribal groups took shape against Dutch colonial machinations followed by heavy-handed Indonesian insensitivity (Webster 2007). Despite their internal diversity, the people of Papua consider themselves distinct from the rest of Indonesia and resent the efforts to resettle Javanese migrants there. Indonesia's Suharto regime in particular exploited, repressed, and even massacred many of the residents of these regions, engendering distrust and inflaming nationalist passions (Anderson 1999). Figure 6.3 shows 2012 protests in the city of Melbourne, Australia aimed towards freeing multiple Papuan political prisoners.

The final type of multinational state occurs where there is no clear majority or dominant nation. In other words, a state is divided into several groups and while one group may have control of the government, there is little cohesion. An obvious fix for this situation is to create a supranational identity that rises above the individual differences. This has been successful in some cases. Tanzania comes to mind as a country that was able to weave each of the over 120 strongly held ethnic identities into a cohesive sense of larger national identity. This is a credit to the first president Julius Nyerere and his

Figure 6.3 Protests to free Papuan prisoners, in Melbourne, Australia
Source: Matt Hrkac from Geelong/Melbourne, Australia, CC BY 2.0, https://creativecommons.org/licenses/by/2.0, via Wikimedia Commons

development of the principles of Ujamaa based on civic principles of communalism (Otiso 2018). Or there can be developed a power-sharing system that provides enough individual national control and coordination at the highest level to make it all work out. The Swiss Confederation provides quite possibly the best example of this, where most authority is relegated to the cantonal level (Bendix 1992).

Unfortunately, many such states struggle to maintain any cohesion at all. These are often the legacy of colonies drawn artificially, superimposing boundaries over existing ethnic and national territories. Sub-Saharan Africa is the continent where this was done most consistently and egregiously. African states were formed without acknowledgment of the myriad ethnic, political, and religious affiliations, and the economies and political system were developed to further European imperialist goals (Carment 2003). Afghanistan, site of so much recent conflict, has long suffered from a lack of any internal cohesion. There are some larger commonalities – explicitly the shared Islamic religion and early government efforts to promote a sense of national culture (Dupree 2002). But ethnic divisions are marked within a mountainous terrain that impedes attempts at cohesion. Many rural Afghans spend their lives just within a single valley.

Afghanistan has long been an artificial creation, inspired more by the colonial powers which hemmed it in as a buffer zone. Even the name of the country was not agreed upon, as separate terms were used to describe various parts of the country (Hyman 2002). The number of ethnic groups range from 50 to 200, divided by language, sect, livelihood and often bound in feudal relations with various clan chiefs who constitute important markers of identity (Schetter 2008). These divisions have never allowed for true nation-building outside

Figure 6.4 Panjshir Valley, Afghanistan
Source: Homayon Khoram

the capital as national ideologies play little role. This is not to say that the country has never been united. Various powers – most recently the Taliban – have at times bound most of the country together by force and repression. In more tolerant conditions, such as existed after the Taliban were removed, politics has aligned with region and ethnicity. The capital has little control over the rest of the country and true state legitimacy does not exist.

Multinational Identification

Even within a region where the bulk of people define themselves as belonging to one nation, there are going to be tremendous variations. These variations exist among separate people, as each individual comes to terms with what national identity means to her. It also varies by region, where perspectives of national identity become enmeshed in each place's political culture. Visions of a nation are going to depend on where one is situated. National identity is produced in the interactions between people who share local practices and have local concerns. Kaufmann (2016) provides the example of how the conception of German nationalism in Hamburg is heavily shaped by that city's seafaring past, where it was one of the key ports of the Medieval Hanseatic League. The city of Pfalz, on the French border, have used their long history of both resisting French aggression and then later collaborating with their French neighbors as an important aspect of their identities as Germans. Kaplan and Häkli (2002) also looked at the significance of borderland national identities and how these differed from how such identities are represented in a nation's core territory.

The idea that national identity is just as much locally/horizontally based as centrally/vertically based has relevance to how we can think of different identities as they exist within multinational countries. Certainly, we must consider the central, official perspective on national identity since this perspective is continuously expressed, even propagandized, in all manner of institutions. But it is also necessary to understand that national identification occurs also at smaller scales. People living within culturally distinct regions form an attachment to that place. If the culture and place are combined, then the attachment to culture mixes with the attachment to region. This is true across the world, in regions large and small. A resident of Brittany, especially if she is culturally Breton, attaches to the Breton region as a culturally defined place. Someone living in Brittany not identified as Breton will attach in another way (Boomgaard 2008).

Under some conditions, this ethno-regional attachment may form into a kind of national identity. But while they experience allegiance to this nation, it does not necessarily preclude identifying with the larger state – which in itself may promulgate national ideals. The tension between smaller and larger forms of identity underlines that people's allegiance crosses many scales – it is "nested" rather than exclusive (Herb and Kaplan 1999). This means that people will identify with smaller and larger regional scales, some of which may be considered nations. It is possible to be both Welsh and British, or Sicilian and Italian. The levels of salience among these identities will depend on whom you ask but an affinity for one does not necessarily block out attachment to the other.

Enabling identification at different scales is important to establishing a multinational state that does not govern through oppression. The civic vision of nationalism, as described in Chapter 2, often corresponds to a larger scale of the state itself and is used as a way to bridge various peoples within a single country under a common identity (Breton 1988). The ethnic vision of nationalism refers to a more circumscribed nation. Depending on the situation and the stakes, people may be more likely to identify with the smaller nation, closing the circle around a shared vision of cultural commonality, or they may be more inclined to consider a larger national bond which may be based on a shared ethos. Yet for a multinational state to function, people must be able to hold on to identity at these two scales. According to Keating (2001), they must be "plurinational" and share more than one national identity. This cuts against Connor's (1972) vision that national identities at a lower scale must be destroyed for strong national identity at the state scale to truly succeed, or that strong identities at smaller scales require political independence. In this sense a British "nation" can exist in the minds of people who also feel allegiance to a Welsh or Scottish nation. As Clark (2000, 275) argues, "the pattern within the British Isles has been the resilience of a diverse and plural system of identities, rather than the rigidity but final shattering of a unitary one." Or people can describe Quebec as a "nation within a united Canada" (quoted in Stojanovic 2011).

To maintain its cohesion, a multinational state must enable an individual to keep two or more identities in her head at the same time. This requires a certain kind of compartmentalization, where different identities serve different goals. It also means that the aspirations of smaller identities must be compatible with the larger state and must be successfully accommodated by the state. How far does the desire for self-determination go: complete sovereignty, independence, some form of autonomy, recognition? In addition, there is always a question of which identity will prevail if a choice needs to be made.

Box 6.1 The Soviet Identity

Figure 6B.1 Soviet identity: money

V. I. Lenin once described the Russian Empire as a "prison of nations" (Weeks 2013). Similar language has since been used to define the Soviet Union. In the case of Soviet identity, "ethnic" or "tribal" identities were often substituted with a common, civic national identity in order to achieve national development (Bekus 2010). Many scholars view the modernization of the Soviet Union as a contributing factor to the spread of Soviet nationalism. For example, modernization changed the social structure of Soviet society by creating new communication networks that allowed the state to reach the entire population (Bekus 2010). Mass popular culture could be seen in newspapers, books, and other media sources for the first time in Russian history that allowed for an ideological reinvention of the nation (Bekus 2010). However, the spread of Soviet ideas through enhanced state communication was not the only contributing factor to the creation of a Soviet identity.

During the 1920s, the Bolsheviks worked on the development of national language and culture, recognizing 169 languages in nationalities across the Soviet empire (Bekus 2010). Bolshevism held the belief that national identity was not essential or permanent, instead it was unavoidable and had to be passed through until a mature, internationalist socialist world came into existence (Bekus 2010). The idea was that national identities had to be embraced, to a certain extent, in order to establish a broader Soviet identity. In some contexts, this strategy worked, such as with younger populations in Ukraine that promoted a unified language and culture to create a strong ethnic identity as Soviet Ukrainians, rather than just Ukrainians (Wojnowski 2015). The hope was that each nationality would not only see themselves as members of their own nation, but also members of the larger Soviet Union.

Figure 6B.2 Soviet identity: stamps

Lenin held a similar view to the Bolsheviks, believing that the right to self-determination would prevent nationalism just as the right to divorce is not an invitation for women to leave their husbands (Bekus 2010). The idea was that if the nations were content in their rights, they would not seek self-determination. However, by the 1930s not every nation in the Soviet Union was satisfied with the few rights they had been granted. Some ethnic groups were eventually considered to be less loyal to the Soviet homeland than others, resulting in repressions and deportations (Wojnowski 2015). The party apparatchiks started using ethnic categories when allocating access to health care, education, and public cultural forums (Wojnowski 2015). More resistance to the Soviet identity and a greater push for self-determination came from these actions rather than the desired embrace of membership to the Soviet Union.

Overall, promotion of a common Soviet national identity, in order to avoid separatism and the development of strong separate national identities, was a legacy of Russian politics stemming from the 1860s (Weeks 2013). While the Soviet Union was the first of the old European multiethnic states to promote national consciousness in response to nationalism, this approach did little to create a solid Soviet identity (Bekus 2010). Soviet leaders changed tactics from promotion to repression, leading to more disdain for the Soviet Union. It is hard to accurately describe what constitutes the Soviet identity as we saw the collapse of the USSR before there was a general consensus for what it meant to be Soviet. The Soviet identity was more of a dream than reality.

The Political Organization of Multinationalism

The political organization of a state can play a significant role in helping to accommodate multinationalism. To what extent are the political and cultural goals of a minority national group recognized by the state? How far will the state go in granting that group an opportunity to make its own rules and govern itself, at least in some spheres? This has been an objective of groups even before the coming of nationalism but became even more pressing once nations aspired to self-determination.

Culture and politics are separate but often require political solutions to enable cultural goals to be met. One issue that is germane to all multicultural societies is the nature of the electoral system (Reynolds and Reilly 1999). The type of system utilized can have the effect of recognizing or even stressing cultural divisions or suppressing these divisions, particularly where a national group does not have a large population. For example, a plurality/majority system, such as is practiced in many English-speaking countries, tends to reward regionally concentrated national groups but suppress groups which are not a majority in any one area. Another system common in Spain, Portugal, the Netherlands, and their former colonies, allows for representation to be proportional to the vote. Groups which are not spatially concentrated then have a greater chance of having their share of elected officials. In countries with potent national divisions, these divisions very often align with particular political parties. The type of system can be a means to enhance representation, depending on how national groups are distributed. In other systems, such as ranked choice, voters list their second, third, and fourth choices. This might be a way to encourage appeals beyond the national group and so soften divisions.

A rubric developed by Mikesell and Murphy (1991) categorized the political/cultural aspirations of minority groups. They defined minority group objectives like recognition, access, and participation. A group may simply seek to be recognized, say with a designation in the national census. Or they may seek something more accessible which could include steps towards more affirmative action policies. Proportional representation or quotas in government offices could be used as a way to broaden participation.

If strategies are not in play as a way to work within the existing state apparatus, there remains the option whereby minority objectives are used to develop alternate polities. Mikesell and Murphy (1991) describe these as either separation, some form of autonomy, or outright independence. To show what these political solutions mean in practice, for example, the Amish population in the United States has achieved a measure of separation and exemption from particular laws. Children are not required to attend school after the 8th grade, for example, and Amish do not have to pay social security or Medicare taxes (McConnell and Hurst 2006). Countries often grant autonomy to political subdivisions. Italy, for example, includes five separate regions/provinces – each of which enjoys particular privileges. The province of Bolzano (Alto-Adige) in far northern Italy provides special protections to the German speakers who are a large linguistic majority and also enjoys considerable fiscal autonomy (Alcock 1992). The final resort of independence has been undertaken in many instances. The most frequent cases have occurred as the result of the dissolution of large multinational states, such as the Soviet Union and Yugoslavia. But there have also been smaller instances, ranging from the relatively peaceful secession of Slovakia to the bloody independence of South Sudan.

The first issue to explore is the degree to which geography is a necessary aspect of the regional solution. Efforts to provide national rights may be accomplished through

Figure 6.5 Amish buggy
Source: Nicholas A. Tonelli

non-territorial or territorial means or, as McRae (1974) termed them, principles of "personality" or principles of "territoriality."

Non-territorial Approaches

Non-territorial solutions have been common through history as separate groups in society were allowed or deprived of certain rights and privileges. Jews within many European countries were singled out as a group and not allowed certain rights. Jim Crow-era laws targeted African Americans' legal rights in non-territorial as well as territorial ways. Non-territorial solutions were also utilized within many multicultural empires (Coakley 2016). The most famous of these was the Ottoman Empire's use of the millet system. Here, every non-Muslim religious community was granted a degree of self-governance in matters of religion as well as civil, fiscal, educational, and penal affairs. They were assessed higher taxes but were exempt from military service (Barkey and Gavrilis 2015). Personality and territorial principles can also be seen in the operation of the Austro-Hungarian Empire. The division between Austrian and Hungarian realms was clearly territorial, with control vested in all but a few governmental functions. The Austrian portion of the empire was very multinational, but the diversity was spread throughout, as most provinces contained a hodge-podge of nationalities. Therefore, the empire accommodated this diversity primarily in a non-territorial manner (Kuzmany 2016). For example, part of the basic law stated that "each people has the inviolable right to maintain and cultivate its nationality and language," including provisions for people to be educated in their own language.

Modern societies have sometimes adopted either a pillared or consociational approach to national minority rights. A pillared society provides each of the recognized minority groups all of the essential resources to function as a complete society within themselves. It provides for the founding of different schools, equal support for religions, and language. In encountering a large immigrant population, the Netherlands once provided each of its constituent populations "integration with preservation of identity" (Prins and Saharso 2010). Lebanon recognized its divisions between Maronite Christians, Sunni Muslims, and Shi'ite Muslims and clearly demarcated each religious group (Calame and Charlesworth 2011).

While a pillared society suggests a country where two or more national groups decide to remain apart, a consociational approach envisions these cultural divisions as a means towards a grand coalition. Along with having a great deal of autonomy over its own affairs, each national group enjoys proportional representation in the country's political system and in its government executive functions, and veto power over major legislation (Lijphart 1979). Such a system requires each national group to trust the other groups, and for the majority to willingly give up some of its power to the minority group or groups. There have been many attempts at creating a consociational democracy, though it would be hard to point to any that are completely successful. Post-apartheid South Africa desired to thread the needle after its discriminatory past and to bring in all aspects of society (Gounden 2010). This involves acknowledging the large racial and national differences and forming a larger umbrella national identity around a single constitution, national anthem, flag, and a shared public culture (Maake 1996).

Territorial Approaches

A territorial approach bestows political/cultural rights depending on where people live. A large multinational state that includes culturally distinct regions might devolve a great deal of authority to all subdivisions. Often it may single out a few as "special" regions based on the cultural peculiarity. The conflation of cultural distinction with territory creates a clearer vision of what the nation may look like. Spatial boundaries contain all of the elements enclosed within and so the process is more efficient (Sack 1986). There is no need to determine who in the population belong to a particular cultural group and so should have certain rights; instead everybody within the region is granted these rights. The association of a nation with a region furthers the prospects for national secession, as provincial boundaries may be envisioned as the boundaries of a new state. Yugoslavia's designation of the Kosovo autonomous region within Serbia beginning in 1945 formed the template for an independent, though not universally recognized, Kosovo state (Tahiri 2018).

The second issue to be addressed is the range of rights which are provided. The option of autonomy for a national region spans a wide range of possibilities from that of an ordinary subdivision to rights which come closest to sovereignty. Political scientists have offered some ways to categorize the degree of self-rule along several dimensions (after Hooghe et al. 2008).

1 How much institutional depth each region possesses and how much its authority may be overridden by the central government. More autonomous regional governments can claim sovereignty over certain functions, without the prospect that these laws will be overturned at a higher level.
2 How much control the nation-region has over government functions like protection, welfare, and education. This could be important in all sorts of policies vital to a region's cultural autonomy. For instance, a region could oversee an educational system operating in a minority language.
3 The ability to tax within a region, and to keep the revenue gained this way, is a vital means of instilling greater fiscal authority.
4 The ability of a region to self-govern – in other words to elect its representatives and its executive branch.

These sorts of prerogatives are significant for all subdivisions, whether culturally divided or not. Individual states within the United States, for example, claim a great deal of self-rule along all four of these dimensions, under powers specified by the 10th Amendment to the Constitution. In other countries, subdivisions enjoy far less power. In Thailand, for example, the governor of each province is appointed by the Ministry of Internal Affairs. The control of education in many countries, like Japan, rests with the central government.

Federal or Unitary Systems for Multinational Governance

In accommodating the presence of more than one nation within a country, there are a few choices available. A unitary state vests most authority within a central government. Subdivisions exist but are rarely provided with special prerogatives and there is no pretense of power sharing between the country's central government and the governments of the regions, at least not hypothetically. A federal system, on the other hand, is designed to ensure the powers of different subdivisions and to share power between the central (federal) and regional governments. Federalism is often contrasted to a unitary government where the central government yields the most power. In this case, a country is made up of a central government and regional units of government, each of which exercises sovereignty within its own sphere(s) and where the regional units are not considered subordinate to the central government (Broschek 2016). In its various manifestations, however, there is a wide range between countries where the central government exerts significant power over the regions and countries where regions enjoy tremendous authority. One rarely found system is a confederation, where each constituent part has more power than the country as a whole. Confederation would be considered a union of sovereign states, each with complete control over its territory and people (Kiljunen 2004). There are few of these. The first government of the United States was a confederation but fell apart under its own internal squabbling and instability. The Commonwealth of Independent States, which succeeded the Soviet Union, was the most prominent. Switzerland and Canada, which call themselves confederations, operate like federal systems, with clear separation of powers (Pinder 2007).

While it may seem that the federal system provides the best option to bring separate nations into a singular state, especially where the nations are arranged geographically, many unitary governments have also been successful in accommodating national differences. The key dynamic is the extent to which the central government empowers a culturally distinct region. In both unitary and federal systems, this has been accomplished through the process of devolution where significant powers are devolved onto a constituent region.

There is also some blurriness here, as states that do not self-describe as "federal" operate as such in key respects. The United Kingdom was once a prime example of a unitary state. While the UK had traditionally recognized some regional distinctiveness among its "kingdoms" of Scotland, Wales, and Northern Ireland (not to mention smaller units like the Isle of Man and the Channel Islands), legislation and administration was vested in Parliament in Westminster, London. This changed in 1998 when the UK began to devolve some powers onto these constituent regions. Each kingdom was provided a regional government that had great control over its internal affairs. McGarry (2012) describes this as asymmetric autonomy. England, representing about 85% of the UK population, was not granted any meaningful home rule, reflecting the lack of interest among the already majority English population. Wales, the kingdom tied most closely to England, was given an assembly and some functions that had belonged to the Welsh Office. Northern Ireland had a separate assembly but also a consortial arrangement with the Irish Republic. Scotland, among all kingdoms, enjoyed the most wide-ranging powers. It was granted discretion over most government functions and had authority over tax rates.

While federalism would seem to lend itself well to cultural diversity, it does not explicitly have to do so. A federal structure does not necessarily reflect major cultural differences between peoples. On the other hand, some very culturally diverse societies may also be federal polities – but these differences are not reflected in the federal structure. A country as culturally diverse as the United States – based on its history of slavery, immigration, and the displacement of indigenous peoples – does not reflect these differences in the federal structure. Australia has also been a federation for its entire history, modeled in many ways after the United States. The Australian states, however, do not align with its indigenous groups. Without a territorial base of power, "first peoples" are compelled to find other means of representation, few of which seem adequate at this time (Cornell 2015). When Canada partitioned the territory of Nunavut from the Northwest Territories, this represented an alignment of territory with a distinct aboriginal cultural group, the Inuit.

Figure 6.6 The Queen at the Scottish Parliament
Source: Adam Elder

Yet many solid examples of multinational federations exist. If national groups are concentrated by region and these national regions roughly correspond with formalized districts, then federalism might provide a level of self-government and a basis for participation in the central government – promoting Lijphart's (1979) consociational principle. This brings up a raft of issues, including how much self-determination is allowed within the region, how well national minority groups are treated outside "their" region (and how well the national majority group is treated within the minority region), the extent to which a minority national region may be continuously overruled and isolated at the federal level, and what is the basis for a shared citizenship in the state as a whole and shared participation in the central government (Pinder 2007, Seymour 2012, Resnick 2012).

Each multinational federal system varies along these key components. One additional aspect to keep in mind is the distinction between how self-determination is applied in the legal documents, notably the constitutional framework, and how it works in practice. Many federal constitutions provide for measures of regional autonomy, up to and including the ability to secede, but these promises are tested and often contradicted in practice.

Spain

Separate kingdoms, each with their own language, were combined to create a larger Spanish nation under the auspices of a king or emperor. From the Spanish Civil War until the death of the longtime dictator, Francisco Franco, Spanish diversity was severely and sometimes brutally repressed. All residents of the country were required to speak in Castilian and displays of cultural difference were put down summarily. With the coming of democracy in the 1970s, Spain recognized regional distinctions with a new constitution. Spain itself is not a formal federation, inasmuch as the terms "federation" or "federal" are found in the 1978

Spanish Constitution, but it operates very much like a federation that legitimizes and funds the self-rule of each region – up to a point (Moreno 2007).

In particular, a few special regions have been developed, with the understanding that these harbor a distinct population: Galicia, Catalonia, and the Basque country. Galicians speak a language closer to Portuguese and also continue a Celtic tradition. Yet, while language pride runs strong, it is primarily a rural phenomenon (Hannum 2020). Much of the truly separatist activity had taken place in the Basque region, where the Euskadi Ta Askatasuna (ETA) fought for independence through terrorist tactics. This group has since been dismantled and Basque aspirations have been carried forward by political means (Álvarez 2017; Mansvelt Beck 2008). Catalonia was traditionally far more subdued, preferring to retain its culture and especially its language through politics. In many respects, it also was the most cosmopolitan of the regional identities, with the major city of Barcelona a center for Catalan language and culture. The situation in Catalonia changed as pro-independence parties took over and supported referenda on independence in 2014 and 2017 (Guibernau 2014).

Ethiopia

Among the most permissive federal structures comes from Ethiopia. Ethiopia contains about 80 ethnic groups and is dominated by four of these groups. No one group represents a majority; the largest group, the Oromo, constitute only one-third of the overall population. In contrast to several African states, which attempted to downplay ethnic divisions as a means of forging stronger national ties, in the 1990s Ethiopia embraced its multinationalism and enshrined it in its federal structure (Habtu 2004). The country divided its nine states along national lines, where five states are given the names of the ethnic group that dominate their territory (Van Der Beken 2019). The alignment between sub-state divisions and ethnic groups is messy, as several groups spill outside the boundaries of their named jurisdiction and members of several smaller groups also share space. However, it does delineate some concurrence of region and nationality and provide a basis for ethno-regional self-determination. According to the 1995 constitution, "Every nation, nationality and people shall have the unrestricted right to self-determination up to secession" (Ethiopian Constitution Article 39). This would be most practical for those largest nationalities but is promised for all.

Constitutional federation has promoted the national claims of the largest ethnic groups. Most significantly, language rights are established in each of the states, especially for education, and ethnic history and identities were further enhanced. Yet several problems have occurred in contradiction to constitutional promises. Given the limited ability to draw regional boundaries, the federal structure does not grant territorial rights to Ethiopia's numerous smaller minorities (Mehari 2018). This becomes a problem also for those larger minority populations that live outside "their" state. Such an arrangement has promoted mobilization along ethnic lines, facilitating the fragmentation of a larger Ethiopian identity and undercutting opportunities for separate groups to work together (Fessha 2017). Finally, the true implementation of ethno-regional powers has not been realized as the central government continues to hold sway (Abbink 2011). The current government (as of the writing of this book) has been especially antagonistic to any strong regional claims (Van Der Beken 2019). In 2021, members of the Tigray nationality have attempted to wrest independence of their province from larger Ethiopia and have engaged in a war with the Ethiopian army.

India

India represents not only the world's largest democracy but also its largest federation. Upon India's independence in 1947 and the development of its constitution in 1950, it faced a tremendous degree of diversity, between major religious groups, some 50 separate languages (and distinct language families), tribal populations, and divisions by caste (Figure 6.7). After the partition of mostly Muslim Pakistan, much of India's distinct geographies have been based around ethnolinguistic groups, which have become the basis of Indian states (Varshney 2013). As 15 languages are dominant in various parts of India, Indian federalism allows for each state to have its own official language. The official language of the center is Hindi and English. Hindi, spoken by about 40% of Indians, dominates in the north. Outside of these regions, English is far less controversial than Hindi, despite its associations with colonialism (Varshney 2013).

The Indian constitution protects linguistic rights but differs from Ethiopia and other multinational federations in its separation of powers. Indian states are given a list of enumerated powers under the constitution. But there is also a loophole that allows the Indian prime minister to remove the government and legislature of any state (Varshney

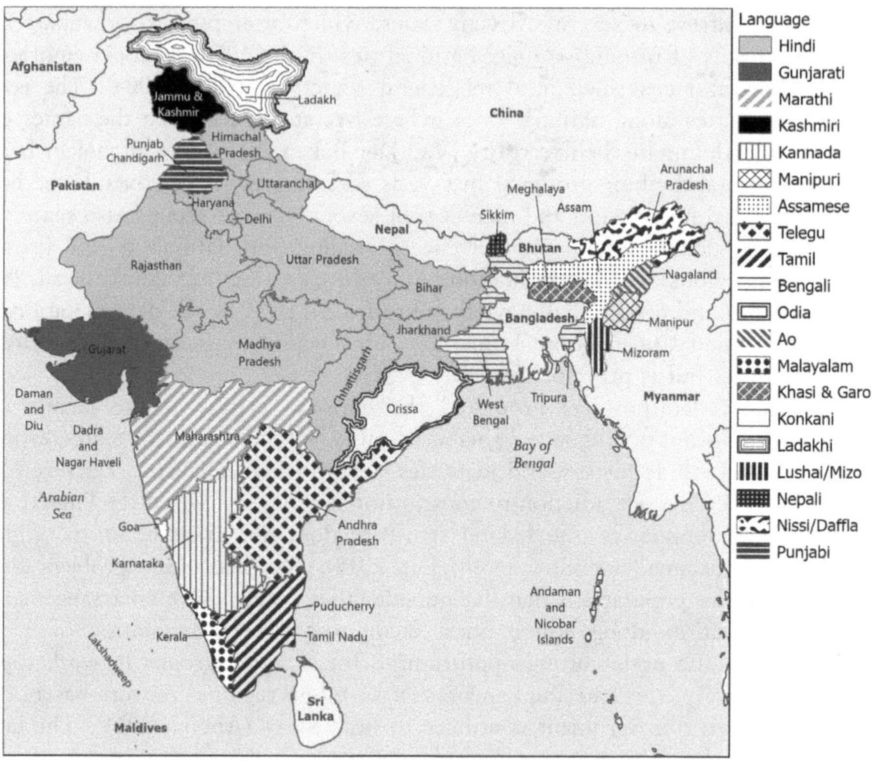

Figure 6.7 Map of Indian languages
Source: Stephanie Burkey

2013). This is no infrequent exercise, as this provision has been invoked more than 100 times! In addition, the constitution prohibits secession and in practice the Indian government has suppressed what it considers secessionist tendencies by force, most recently when the government put the entire state of Jammu and Kashmir under lockdown after dissolving its special status (Bhargava 2012, Yasir et al. 2019), leading to global protests. As with Ethiopia and other countries with numerous linguistic minorities, the separation into linguistically oriented states does not often bode well for smaller minorities without access to power (Bhargava 2012). On balance, the system of federation has done a great deal to quell ethnic acrimony in such a diverse country, yet the government employs a fairly heavy hand against regions that are considered too assertive.

Supranationalism in Concept and Practice

We have discussed the nesting of several political loyalties played out between the state and identities manifested at smaller scales. But how does it work when considering identities at a scale larger than that of the state? Supranationalism considers these nested loyalties and divisions of powers. Rather than moving from the state down to the constituent national parts, it looks upward from individual states to some larger entity.

Supranationalism is distinct from regular international organizations, based on common trade or defense interests. It employs several mechanisms which bind together those states within such a supranational union. The term itself has gained a great deal of currency with the accession of the European Union (EU), considered the best example of a supranational organization. Undoubtedly this means that most discussions of supranationalism are intertwined with the complexities of how the EU is governed. Yet supranationalism needs to be considered as a more abstract concept, since its attributes can apply to several political entities.

It may be easiest to divide the concept of supranationalism from its implementation. The term supranationalism has only been in existence since the late 1940s, when Albert Einstein discussed the desirability of corralling national states into a more peaceful union. The British Prime Minister, Winston Churchill, also weighed in, encouraging war-ravaged European countries to find a way to band together meaningfully and permanently – though it is doubtful he saw Britain as part of this arrangement (Mauter 1998). Later the French foreign minister Robert Schuman articulated its value: "The supranational is situated equidistant from, on one hand, international individualism and, on the other hand, a federation of states subordinating themselves to a superstate" (quoted in Kiljunen 2004, 20).

All federal governments subscribe to the notion of the sovereignty of the federal state itself. Sub-state entities – cantons, regions, provinces, states – enjoy several distinct functions and participation in the central government, but they cannot override the sovereignty of the federal state itself. Citizenship is likewise conferred at the federal level – residents of Quebec (a province with strong powers in a multinational federation) do not carry Quebec passports but Canadian passports. The federal government will issue currency and conduct foreign relations (Kiljunen 2004).

Confederation brings together completely independent and sovereign states into a union. All decision making has to be agreed upon by each state; both citizenship and identification is embodied in the individual states as well. Everything must be negotiated and arbitrated. It is easy to see why such political arrangements have not lasted too long.

Supranationalism combines aspects of both federation and confederation. For some functions, the supranational union enjoys supremacy, whereas the member states maintain their sovereignty over other functions. Passports can be issued by the supranational union, there is an overarching legal system, and a supranational government can make legislation without the necessary agreement of each constituent state (Kiljunen 2004).

An alternative aspect of supranationalism could be thought of as a supranational imagination. Two examples illustrate this. The independence of the Spanish colonies in South and Central America was part of a larger movement that occurred at the same time. Each of the newly independent states shared common attributes, notably language, religion, a common colonial heritage, and the exclusion of most indigenous communities from political life (Lynch 1973). Simon Bolivar is most associated with each of these independence movements and the successor states that emerged. It was his vision that propelled many of these societies to independence and his aspiration for each to join in some sort of union. This would form a cohesive framework and be provided with a measure of sovereignty to complement that of each of the constituent "sister" states (Collier, 1983).

A more recent example involves the supranationalism of the Arab world (Culcasi 2011). As with the South American Spanish-speaking states, the Arab world constitutes many countries aligned by the common religion of Sunni Islam and the Arab language. The idea of a broader Arab entity, transcending the various Arab states, has been a significant factor since World War I and the dismantling of the Ottoman Empire. This vision has long been reflected in maps, with titles like the "Arab Homeland" and thick boundaries delineating this space. While the boundaries shift over the years, the notion of Arab supranationalism remains (Culcasi 2011). Politically, the creation of a pan-Arab structure is most notable in the Arab League, though there have been attempts at much tighter structures, such as the United Arab Republic which briefly conjoined Egypt and Syria.

The European Union as a Supranational Entity

Supranationalism is embodied in the development of the European Union, which has now become the most politically and economically articulated form of supranationalism. This would seem a remarkable choice since, unlike Spanish America or the Arab World, "Europe" is enormously heterogeneous. It includes a hodge-podge of languages, many of which have come to diffuse well beyond Europe itself. While mostly Christian, Europe hosts many denominations that have long been in conflict. Both World Wars were initiated by countries within Europe, fanned by long-festering tensions. In that regard, there would seem to be scant reason such a diverse and long-warring set of countries would emerge within 50 years to represent the iconic example of supranational cooperation.

The European Union began as a customs union in the 1950s between just a few countries, related to trade in coal and steel. By the 1960s, it had evolved into the European Economic Community or Common Market, which eliminated tariffs between member states. The 1970s and 1980s saw the expansion of the Common Market to more countries and discussions of how to further integrate this consolidated economic zone and to add several political structures. Moreover, the Schengen Agreement allowed for the removal of border checkpoints between most countries. In the 1990s, the Maastricht Treaty created the modern European Union, and a truly supranational state. Common citizenship, freedom of movement, a common currency, and a central European government were all hallmarks of this new structure while, at the same time, member states retained their sovereignty (Mason 2008).

The European Union showcases the potential for dividing sovereignty across geographical scales. The member states of the EU guard their rights and powers. Only about one percent of each country's GDP goes to EU administration and governance, a small fraction of what is spent by each national government (Forslid and Nyberg 2020). The interplay between European identity and that of the member states is a long-running process, with attempts to create a sense of unity out of diversity. According to Forsyth (2007) this was already a work in progress 15 years ago. The mechanisms of the European Union had far outstripped the sense of a shared European identity.

Given the differences between individual countries, the European Union has succeeded in creating a sense of shared identity and of trust (European Union 2021; Forslid and Nyberg 2020) (see Figure 6.8). A sense of cultural identification is aided by Europeans' self-perception as the natural successor region to the ancient Greek, and so the carrier of civilization (Agnew 2002). Most residents on the continent share a strong and positive identification with the EU that has not wavered, with a median of 72% of adults in EU countries having a favorable opinion, almost three times as many as have an unfavorable view (Fagan and Gubbala 2022). What is more, citizens on average trust the EU more than their own domestic government. A sense of patriotism, which is defined as affection for one's own nation and tends to promote civic mindedness, is quite compatible with broader support for the EU (Huddy et al. 2021). Politically, the EU has settled into a meaningful division of powers between itself and each member state. This principle of subsidiarity allows for some standardization but retains the peculiarities of each member state (Forslid and Nyberg 2020).

At the same time, there are differences in attitudes towards the EU by ideology. With the notable exception of Greece, those on the ideological left tend to have a markedly more favorable view of the EU than those who identify as politically right. This ranges from a small gap in Germany to a giant chasm in Hungary (Fagan and Gubbala 2022). And because ideological leanings in Europe increasingly map onto place of residence – with cities much more liberal than the surrounding countryside (Kenny and Luca 2021) – this means that the European Union is perceived less favorably in rural areas.

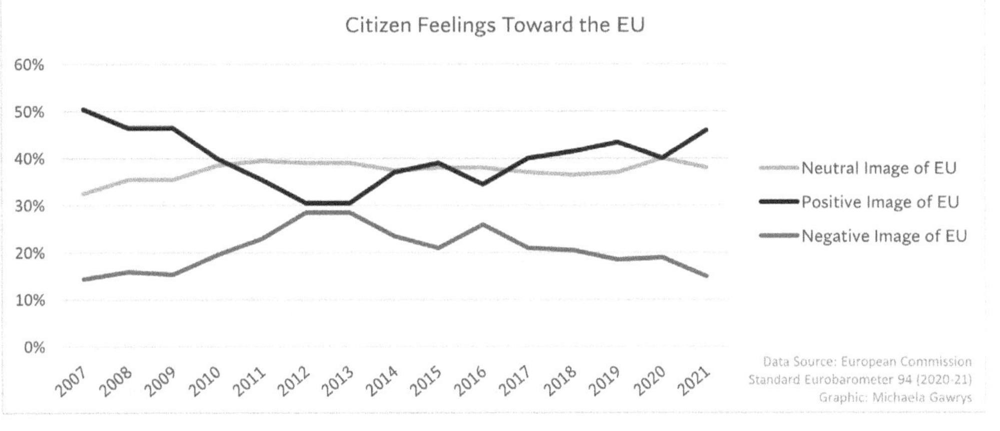

Figure 6.8 Citizen feelings towards the EU
Source: Michaela Gawrys

If anything, the last 15 years have provided more challenges to a larger European identity. Europe was very much rattled by the currency crises, which saw stronger economies such as Germany asked to assist the economies of Greece, and to a lesser extent Italy and Spain. That all countries within the Eurozone, the major European achievement of the 21st century, were tethered to a single currency eliminated much of the flexibility of shifting exchange rates. In part, Greece's experiences with the Eurozone during the fiscal crisis activated more jaundiced views (Tsirbas and Sotiropoulos 2016).

The United Kingdom's decision to exit the EU altogether threatens the dismemberment of the entire supranational enterprise, though enough structures had been put into place that secession is still a long and complicated path. It does highlight many weaknesses in the EU but also reveals that the UK was a unique case. It had long been the most Euro-skeptical country, refusing to go along with open borders and common currency. In fact, Forslid and Nyberg (2020) argue that the difficulty in Britain's exit, where the EU simply refused to allow the UK continued access to its trading bloc, underlined just how difficult leaving this now mature supranational entity could be.

Conclusion

The predominant political ideology of the nation assumes a homogeneous identity that has the right to self-govern. This right to national self-determination appears in international charters and legal doctrine and is a major contributor to national conflict and minority strife. This is because a true self-governing and homogenous nation, a nation-state, is nearly impossible to achieve. In reality nations contain all sorts of heterogeneity, be they ethnic, religious differences, or radically different political views. In practice, nations either work to accommodate these differences, or differences are suppressed. This chapter has explored the many ways in which states in search of cohesive national identities have dealt with this reality.

Most nations are embedded within multinational states and exist as "sub" state nations. Sub-state nations exist politically in a variety of ways. Depending on the makeup of states, nations can clearly dominate states as majorities like in Japan, be given devolved territorial power as substantive regional entities within states such as in Catalonia or French-Canadian Quebec, or be one of multiple nations within a state of no clear territorial or demographic majority. This final type of multinational state can be stable, as is the case of Switzerland, or politically tumultuous as an overarching civic identity may not yet exist.

States employ a variety of strategies to accommodate for or repress multinationalism. Some strategies are non-territorial, marking every member of an "other" group as distinct in some way. These can be exclusionary and punitive, such as in Jim Crow-era America or apartheid South Africa. This distinction can also be used to demarcate cultural privileges, like the ability to wear a distinctive garment or attend a school that speaks a minority language. Some of these result in national cultures being distinct and separate from the majority nation, while others result in a political coalition with strong national distinctiveness.

Contrary to prior political belief, maintenance of more than one scalar national identity is now understood as necessary for civil cohesion. Multiple identities can be complimentary and serve multiple goals for individuals and polities. A plurality of nations, cultures, and identities has led to multiple examples of strong states.

Chapter 7

Ethno-Regionalist and Ethno-Nationalist Movements

The previous chapter discussed how cultural diversity can be accommodated within a single country. A multinational state focuses its claims on a shared destiny which transcends ethnicity. There is a nationalist basis for these sentiments as well, but they either point to a larger identity – e.g., "British" as opposed to "English" or "Welsh" – a shared legacy, or a civic conception of the nation. Since most states in the world are effectively multinational, this larger identity underlies most successful polities. If it starts to fray, especially at the edges, then ethno-regional identities grow stronger and more assertive. History is replete with cases where such an arrangement is viewed as untenable. This is especially true for nations out of power who are desperate for self-determination. They seek their own sovereignty and a clean break from other nations they have been forced to live amongst.

A larger sense of identity fights against a more circumscribed sense of ethnic identity that marks this group off ethnically and considers the state it controls to be more of an extension of its culture. Breakaway ethno-regionalist movements believe that they need to be able to express their national identity within their own political state. It also implies (or sometimes states explicitly) that the proper political place for a nation lies within a polity oriented around shared ethnicity and away from multinationalism. To these new nationalists, diversity seems like it has been a raw deal, where they have been stifled in their own goals. In this regard, ethnic-based movements provide a marked contrast with the sort of multinational and supranational entities discussed in the last chapter.

How Culture and Region Interact

The desire to break away and create a separate nation-state occurs as a reaction to an existing situation of multiculturalism or multinationalism. To understand the choices and motivations available to would-be separatists, it helps to see how countries become diverse in the first place and how this intersects with issues of territory and scale.

The first way occurs as a result of migration from other places. Unless completely sealed off, like North Korea, most countries experience some immigration. Some groups move into a country by force, especially evident in the brutal history of slavery. Others come as a result of turmoil or persecution at home, entering as political refugees. Still others decide to arrive voluntarily, as a means of improving their economic fortunes, to increase human capital, or through marriage. Immigration will affect most countries which open their borders a little or a lot. It will affect even those highly restricted countries, as migrants cross the border illegally. Historically, immigration affects overseas

DOI: 10.4324/9780429435782-7

colonies, as migrants were brought in, forcefully or voluntarily, from many different lands. These migration patterns result in greater diversity, as separate groups mixed with one other.

While the first course of diversity can and does create a host of microgeographies throughout the urban and rural landscapes (see Conzen 1993), it does not often lend itself to the sort of ethno-regionalism which could produce strong nationalist sentiments. There are practical considerations why this is true. The most important is related to issues of scale. As immigrants come to a country and then sort themselves out – either through desire or compulsion – into separate neighborhoods or communities, they often concentrate into particular areas. Yet these areas are scattered, creating small islands or archipelagos of ethnic predominance. Even in those regions that have witnessed strong population growth and ethnic predominance, such as the Cuban population in southern Florida, the smaller geographies of an ethnic enclave do not elevate into the consolidated space of a national region. In addition, most of these microgeographies are recent and subject to shifts as the ethnic character of different communities, neighborhoods, and rural spaces shifts, as is common over time. Strong ties develop to specific places, and sometimes these ties are multigenerational, but they rarely reflect a legacy going back centuries.

The second pathway to diversity comes from the expansion of a state's territory to incorporate adjacent lands. Through the process of conquest or marriage and annexation, the state incorporates culturally distinct places, forcing separate peoples and territories under its control. Before the modern state system was established, the boundaries of separate countries were not so stable. Expansion and retraction were commonplace. Stronger countries would take over adjacent lands or an advantageous marriage united together two kingdoms. The strongest countries would themselves become empires, composed of many cultures, but other states expanded as well. With the adoption of the system of territorial states, and the acknowledgment of sovereignty and agreed-upon borders, territorial changes became more remarkable, as a result of major wars or imperial collapse (Murphy 1996).

While the first pathway to diversity creates a host of obstacles to successful national separatist movements, given the right set of conditions, ethno-regional diversity presents conditions often favorable to enhancing national identity and calls for an independent state (see Williams 1982). In such instances, distinct cultural groups have long occupied the land, consider it to be their homeland, and manifest cultural markers – language and religion most prominently, but also building styles, cuisine, dress, and all of the things that constitute culture. Moreover, these areas are likely to be numerically dominated by members of the distinct group, who may occupy that region fairly exclusively or share it with several minority populations. Finally, in most cases, the lands in question are contiguous and substantial enough to form the basis of a new country. While not unheard of, regions which are geographically fragmented rarely form the basis of a new polity.

Most countries contain both types of diversity. They have experienced significant immigration or forced migration while encompassing lands that are culturally distinct. Numerous waves of immigrants from Europe as well as receiving slave ships from Africa further populated the New World countries, displacing Indigenous peoples. Many of these tribal areas were eradicated, with people within them either killed, forcefully removed, or pushed into an unequal incorporation with the European settlers. In other cases, a once large culture area was reduced to a remnant reservation. This mixture of diversities is also found in many European countries. Traditionally, these countries comprised culturally distinct regions from long histories of state consolidation, imperial dissolution, and the ebb and flow of borders. While such countries also hosted migrants, notably Roma and Jewish peoples, in the last 60 years they experienced mass migration from Eastern Europe and the Global South. How to manage both sorts of diversity is at the top of each country's public agenda, as grappling with so-called "special regions" and the rising tide of immigration creates all sorts of avenues for potential conflict.

Secession Motivations and Tradeoffs

We can separate any desire for greater autonomy into rights-based and goal-based (Caney 1998). What is the moral justification for this and then what outcomes would a national minority like to see? The rights of a nation to secede have been propounded since the time of Rousseau, but the question is who are the nation? Rarely do citizens of a region speak in a unified voice, and even if there is a strong inclination to secede there is also a question as to whether outsiders, even those other members of the state, should have some say in such a dramatic change. Political philosophy has varied on whether secession should be a unilateral right or a circumstance that occurs only in the most restricted instances. The secession of every group who wanted their own independent country would lead to international anarchy, as the boundaries and integrity of the existing state would always be in doubt (Freeman 1998). The global order is likely to be weighted against such efforts, even if done under the most favorable of situations. And the situations are rarely clean and favorable. In many cases, there are other peoples as discussed above – they should have an opportunity to remain in the state. The state itself can be injured by the loss of its territory as this will deplete its resources, disrupt integration, and compromise territorial integrity through the formation of enclaves and exclaves.

All of this tends to disfavor outright independence. However, the spectrum of possibilities ranges from absolutely no accommodations or special treatment to complete independence (Mikesell and Murphy 1991; Griffiths 2016). There are many variations within each of these possibilities. There might be the benign neglect where a group is offered no special treatment or those cases where members of a group are actively persecuted, such as is experienced today by groups such as the Rohingya in Myanmar and has been a sad part for many groups throughout history. Even the terms of complete independence can vary. The dissolution of the British Empire was followed with the creation of a British Commonwealth, an organization where the Queen of England remains the titular head, but which does not stand in the way of sovereignty (Bull 1959).

True independence is often not even the end goal for many advocates of so-called separatist or nationalist movements (see Lehning 1998). This has been clearly demonstrated in many of those regions which are run by avowedly nationalist groups – such as Scotland and Quebec – but where the independence option loses to the status quo when that option is put to the vote. Depending on where these regionalist/nationalist movements begin, there may be some simple aspects of recognition, for instance the right to use their own language in a number of situations, which can satisfy many members of the minority nation. Stepped up from this might be demands for the region to have a greater say in its own governance, with decisions as to which rights belong to the region and which to the larger state. Nationalist groups may seek some sort of proportional representation in the state congress or parliament or develop the sort of political parties which achieve much the same result. For example, the Scottish Nationalist Party predominates in Scotland but also represents a distinct bloc of votes in the British Parliament as a whole. The Parti Quebecois in Canada performs the same function, with frequent control over Quebec's provincial government and a bloc of members of parliament at the federal level. Both such Scottish and Quebec parties are often the third largest political parties in the entire state. Some countries, like Belgium, have only regional-national parties (Mulle 2016). At the same time, for some nationalist groups, independence is by far the preferred option among most of the people.

History has been littered with failed attempts at secession by nationalist regions. According to Hechter (1992), success is predicated on a mobilized population within a territory deciding that secession is in their best interests and that the overarching state's leadership sees the net cost of ceding this region as negative. In other words, they make the rational choice that keeping a restive region is more trouble than it is worth. This perspective also lends itself to a cost-benefit framework (Bartkus 1999) which applies to both the region and its people and the overarching state and its leadership. If the benefits for both region and state exceed the costs, then secession makes a great deal of sense. If the costs to the region are too high, then demands for secession will fall flat. If the overarching state foresees higher costs, it will do everything in its power to impede the separation of a given region. "Costs" must be seen as a fluid, multifarious concept that embraces economic, cultural, military, political costs – but also extends to pride and attachment to the region itself.

A very weak state may not have a choice in preventing secession movements. The case of Yugoslavia in the 1980s, following the death of Josip Broz Tito, betrayed many of the fissures that were causing the country to unravel. Yugoslavia was beset by massive public debt, disaffected job seekers in the major cities, a divided leadership – all in addition to the multinational state cobbled together after World War I (D'Souza 1994). However, there is the general principle that sovereign states should be allowed to maintain their boundaries as recognized by other countries and by international bodies such as the United Nations (Freeman 1998). In practice, this translates into disapproval of any country meddling into what was seen as another country's internal affairs. Outside countries often withhold recognition of successful separatists until the state they separated from recognizes it (Kosovo).

The enumeration of costs and benefits for a region is useful, though any rational reckoning might be swamped by strong nationalist emotions. The basic hierarchy would follow from whether a group and/or its culture feel as if they are in mortal threat of extinction. Clearly a sense of active persecution would prompt a desire to secede. But cultural decline as the region is overwhelmed by the majority culture would also prompt urgent concern. A region may feel as if they do not come off so well fiscally as the state redistributes their resources to other, poorer regions. A region may feel as if they are forced to compromise on some cherished principle and independence is the only way to break this. Finally, they may simply want no business with the politics of the larger state, feeling their voice is not heard.

The costs of independence are just as varied. The main cost occurs when either the people in the region are not united and/or when state resistance is strong. The lack of unity among all potential stakeholders in the region can stymie the potential for greater autonomy. Other potential costs are more utilitarian but must be considered. A poor region without adequate resources may find independence a drain on prosperity. Losing a financial lifeline could devastate its population. Further economic costs, especially where the secession is unwelcomed by the state, could entail the closing of borders to people and trade, thus choking the lifeblood of the new nation. The obverse is that well-off regions have a greater incentive to secede. However, the economic advantages and disadvantages are often played to conform to the narrative of would-be separatists.

Nationalism within Fragmented Geographies

At the beginning of this chapter, we discussed the two pathways towards greater cultural diversity within a country. The first pathway, via immigration and forced migration, profoundly shapes a society, and alters how it views itself. It also produces an array of microgeographies. Some groups assimilate, while others are fairly segregated, either on their own terms or because of societal prejudice and discrimination. And every society contains groups who have been kept apart, segregated socially and spatially. This elicits a reaction from members of the groups themselves, unwilling to continue in this marginal position and looking for political solutions.

The proposed solutions have at times included greater levels of autonomy, or even outright independence. Non-territorial solutions have been tried as some countries have experimented with the issuance of group rights. This allows the group the opportunity to participate in society as a bloc and may allocate resources and decision-making for aspects affecting the group to designated leaders (Fincher et al. 2014)

Territorial solutions are more elusive. Without a large, contiguous territory, there is no material to be claimed as a prospective new state. Another impediment lies in the lack of a homeland. Scattered groups may not have a clear homeland or a region for which they form some historical attachment. This is especially true when cultural groups have been ripped away from their homes and all ties are dismembered. Groups that can overcome both of these impediments may be successful, but it is a difficult path.

In those cases where an ethnic population feels itself to be sufficiently apart to constitute an independent nation, we might consider it to be something of a diaspora (see Chapter 3). Diasporas constitute a network between members of a nation who live outside of the national territory, but they can also expand to include those groups who have no national territory to call their own (Cohen 1996). Members of a diaspora enter a society, occupy communities, and build lives apart from the dominant population. Yet part of their challenge is in how to develop out of this a sense of national identity and in addition provide the ability to mobilize the members of the nation into demands for a political solution.

Jewish Fragmentation and Zionism

The classic example of a successful attempt is that of the Jewish population in Europe. While Jews trace their origins to the "Holy Land," for centuries they had little control over that space (Safran 2005). They were forced instead into societies where they were grudgingly accepted, blatantly persecuted, run off, or killed. The microgeographies of settlement resulted from Jews being forced into ghettos or restricted to certain villages. This spatial separation was accompanied by economic and political restrictions that forced them into particular occupations and often prevented certain rights, such as the ability to own land.

The conditions in a city such as Metz, France were typical. In the 1780s, Metz had a comparatively large Jewish community, welcoming Jews who had been expelled from other French towns. Metz also served as a spiritual capital for Jewry. Yet here, in a comparatively tolerant area, Jews were forced to live in a tiny ghetto on the riverbanks which was locked at night. They were forced to wear a yellow star, a widespread practice later revived by the Nazis. And they paid lots of extra taxes, effectively financing the entire local economy (Goldfarb 2009).

The 19th century was a period in which many Jews were enfranchised and allowed to become full citizens, but the precarity never went away. Whenever tough times came, Jews were often the scapegoats. Enfranchisement could turn into vicious persecution, sometimes within a matter of years. The roots of modern anti-Semitism, from notions of racial purity to the scapegoating of Jewish people, laid bare how difficult living in Europe could be. In the early 20th century, anti-Semitism increased dramatically in Eastern Europe, the Russian Empire, and even in Western countries like France, where the Dreyfus Affair (in which a Jewish officer was unjustly convicted of espionage) reminded Jews of the threat throughout Europe (Beller 2015).

The political solution to this was via the ideology of Zionism. There was no possibility of carving out territory from the existing archipelago of Jewish communities and neighborhoods. And while some political leaders, many of them non-Jewish, had suggested alternative spaces, the Jewish diaspora embraced Zionism to advocate a return to the Middle East, specifically Palestine (Azaryahu and Golan 2004). This was based on the longstanding idea of a "return to Zion" (Gitelman 1998). The ideology of Zionism was articulated most prominently by Theodore Herzl. Herzl posed his Zionist ideas within an explicitly nationalist framework and argued that the only reasonable homeland for Jews was in Palestine, then a part of the sprawling Ottoman Empire (Golan 2008).

The political recognition of Zionism came slowly. The Balfour Declaration of 1917 affirmed the British intention to establish a home for Jews in Palestine, an area in the last gasps of Ottoman Empire control and soon to be a British Mandate. Israel was established in 1948, after the Holocaust tragically demonstrated to all the danger for Jews in relying on any given political situation. An independent Israel allowed Jews from all sorts of backgrounds a place to go and led to the introduction of many different populations, including a large number of Arab Israelis. From its inception, Zionism was always conflicted between a more political and civic-minded nationalism and a more ethnic-oriented nationalism (Conforti 2010). The movements of Jews from so many distinct contexts continues to spark tensions in how modern Israeli nationalism is defined. The older, more unified idea of what Israel means to its people has been obliterated, as many groups offer their own conceptions to the mix.

This is also true among the broader Jewish diaspora as Israel only contains a fraction of the global population. Most of Israel's population were effectively refugees, taking advantage of "Aliyah" or the right of return among all Jews. Jews in more comfortable conditions remained in place. Only a minority of American or British Jews feel highly attached to Israel, and few visit the country or learn Hebrew (Gitelman 1998). However, Israel's existence strengthens Jewish identity. Even if Jewish people do not go to Israel, they know that they can and that there is a place where they will be free from persecution. Despite the concerns many Western Jews have with Israeli politics, especially their relationship to the Palestinian people, the ties between the Jewish diaspora and the Israeli homeland are quite powerful (Safran 2005).

Black Nationalism in the United States

The situation of many Black Americans spurred some to consider a nationalist solution. African Americans long suffered involuntary servitude, Jim Crow-era persecution, a history of discrimination and the denial of full standing in US society. The prospect of continuing under such conditions has made Black territorial nationalism more attractive. Yet the Black population grew more fragmented with large-scale internal migration from the rural south to the urban north in the 20th century. This, accompanied by immigration from several Caribbean countries, made a territorial solution unlikely.

In contrast to European Jews, there was never a commonly acknowledged homeland for Black Americans. It was not for a lack of suggestions. Many white Americans in the ante-bellum period proposed that, rather than remain in the United States, free Blacks should instead go elsewhere. The American Colonization Society rose to prominence by arguing that free Blacks could end up subverting white authority and becoming a burden to American society. Therefore, they should be resettled somewhere else. This was accompanied by a push from many American states which looked for somewhere to place manumitted slaves and also significant populations of free Blacks (Whyte 2016). To fulfill their intent, they bought land in western Africa and started to settle Black Americans there. This colony was named Liberia, and this emerged as the focal point of recolonization. Liberia became an independent country in 1847, with the capital named after a former US President. It also proved to be a uniquely dangerous place for those Blacks who migrated there, and many died under extraordinarily harsh conditions (McDaniel 1992).

The notion of a Liberian homeland was quite contentious among whites and Blacks. Many, like the author Sarah Hale, claimed that only in Liberia could Blacks become truly civilized (Ryan 1995). Even many abolitionists, such as William Lloyd Garrison and Harriet Beecher Stowe, argued initially for this. The focus was the eradication of slavery followed by the removal of manumitted Blacks to a territory in Africa where they could form a free society. But for many Blacks, the idea of a Liberian homeland or any other extra-American territory negated the sustained experience of Black Americans in the United States. Frederick Douglass (among the most famous Black Americans of the mid-19th century) stated:

> we are here, and we are likely to remain … We have grown up with this republic, and I see nothing in her character, or even in the character of the American people as yet, which compels the belief that we must leave the United States.
>
> (quoted in Ryan 1995)

With a few exceptions, such as Marcus Garvey's "Back to Africa" movement in the 1920s, most Black Americans agreed with Douglass and considered the US to be their country, as racist as it may be. Instead, they sought ways to build a better life and for most, the Great Migration, from the rural south to the urban north, was the pathway to better opportunity. Unfortunately, Black migrants were often met with the same racist pushback they had experienced before. For many the frustration of segregation and discrimination led to a push for some form of territorial nationalism. In distinction to the early 19th-century push to Liberia, Black people primarily directed this new nationalism themselves. Some looked at the notion of creating a Black country within the borders of the United States itself. Harry Haywood argued in 1948 that African Americans were in

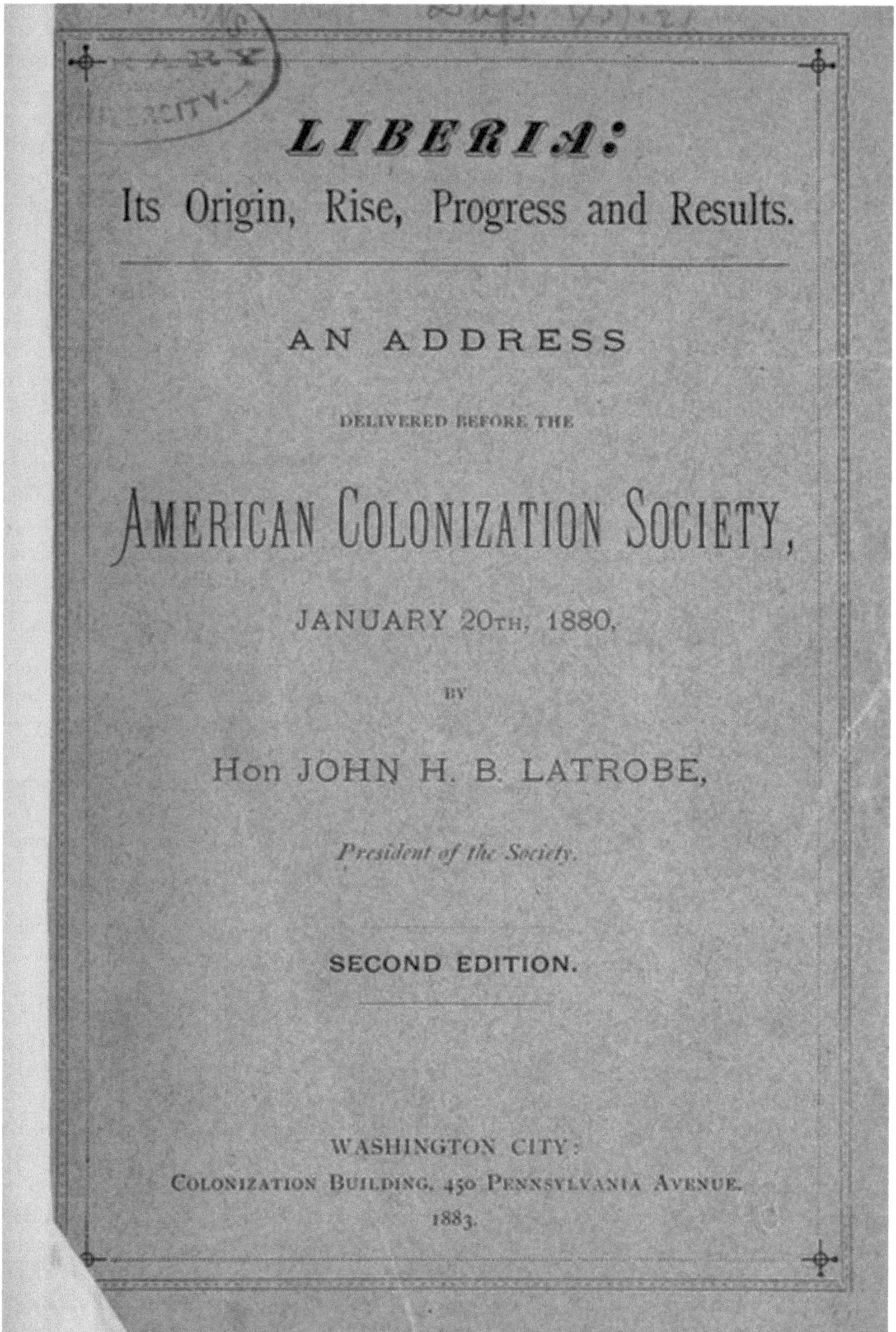

LIBERIA:

Its Origin, Rise, Progress and Results.

AN ADDRESS

DELIVERED BEFORE THE

AMERICAN COLONIZATION SOCIETY,

JANUARY 20TH, 1880,

BY

Hon JOHN H. B. LATROBE,

President of the Society.

SECOND EDITION.

WASHINGTON CITY:

COLONIZATION BUILDING, 450 PENNSYLVANIA AVENUE.

1883.

Figure 7.1 American Colonization Society Pamphlet
Source: Library of Congress

fact an "oppressed nation" with their own institutions, a common English vernacular, and concentration in the Black Belt of the American South. Nationalism could be seen as a remedy for the oppression that Blacks experienced (Valls 2010).

But as Scott (2012) argues, this larger territorial vision never really predominated. Most later 20th-century Black movements were far more concerned with instigating societal transformation at the community scale. Rather than the idea of a true separate territory, impractical to fulfill, elements of nationalism survived in the form of "community nationalism" that aspires to much greater control over the institutions that service Black neighborhoods (Valls 2010).

Indigenous Nationalism in Australia

Indigenous nationalism is quite different from the Jewish or Black cases discussed above. Rather than stemming from new groups who entered a region, Indigenous groups found themselves displaced in their native homeland. In that sense it is related to national diversity resulting from conquest but differs in that this stems from settler colonialism, in which peoples from the conquering society take over the land and begin to occupy it, moving aside those who lived there before, often for centuries.

Settler colonialism and the displacement of Indigenous populations, like the removal of various Celtic tribes in classical times, was a factor in the expansion of the great empires such as the Roman and the Han. Yet when Europeans took over the entirety of the Americas, Australia, and New Zealand, they accelerated the settlement of vast territories on a scale never seen before. In part, this was due to the decimations caused by the introduction of Old World diseases to vulnerable peoples, causing Indigenous populations to crater before any formal conflict ensued (McNeill 1998). For most Europeans, these lands were virgin territory needing to be tamed and cultivated; the destruction of Indigenous cultures and lifestyles could only expedite this process. In each of the large British colonies turned countries – United States, Canada, Australia, New Zealand – the rich tapestry of native life was torn apart as British settlers and later European immigrants continually pushed through continental interiors. In some cases, native tribes were eradicated, in other instances the populations were displaced, sent along to undesirable lands, or forced to make their way to the outskirts of settler society (Cronon 2011). This resulted in a series of dependent spaces at smaller scales.

The development of Indigenous national identity must overcome several obstacles. The attachment to a homeland is clear as Indigenous groups have been in place longer than the original colonial settlers or any future immigrants. They have greater claim to the land being *their* land which provides them greater standing (Cornell 2015). Yet the difficulty has been in retaining rights to land that has been steadily whittled away by land-hungry settlers. The other impediment comes from trying to create a more consolidated identity that supersedes regional tribal identities. In this respect, the challenge for Indigenous groups is similar to that of myriad ethnicities stuffed into a single ex-colonial territory.

The aboriginal populations of Australia have had a particularly challenging time achieving any formal political autonomy, even when compared to American Indians or Canadian First Nations (Bradfield 2006). Much of this can be traced to how Australian identity has spurned any sort of group identity or what they view as a "separatist" agenda (Bradfield 2006). Most of the organizations designed to promote Indigenous rights have been developed on white terms. Even when compared to the woeful history of

North American Indigenous relations, Australia's government was particularly hostile to Indigenous institutions; they refused to consider Australian Aboriginal people as any political entity (Cornell 2015).

Few aboriginal peoples in Australia desire classic independence but there is widespread desire for much greater political autonomy. The Indigenous groups have broken through Australian reluctance to see them as political groups with standing and have functioned as distinct nations with their own set of institutions – not simply institutions imposed by the federal authorities (Cornell 2015). The scale of self-determination is also important here. For instance, Australia encouraged the development of aboriginal outstations, where kinship communities are established on meaningful pieces of land, allowing very small groups to practice some form of self-determination (Peterson and Myers 2016). These outstations are not self-sufficient, however, relying on some financial support from the Australian government, and they are extremely local, averaging between 100–200 individuals apiece (Palmer 2016). Creating a larger polity, with representation in parliament, has proven to be much more difficult, though several aboriginal groups have decided to simply focus on building up their own autonomy, apart from the state (Maddison 2020). Some Australian Indigenous groups have attempted to create much broader and comprehensive nations (such as the Nonegar Nation) out of more dispersed groups (Cornell 2015).

Box 7.1 Nunavut Identity

Figure 7B.1 Nanavut identity
Source: BaShildy, CC BY 3.0, https://creativecommons.org/licenses/by/3.0, via Wikimedia Commons

The "Nunavut identity" in Northeastern Canada evolved from a primarily tribal identification to a larger, regional identification. Scholars believe that there were close to 20 Inuit traditional tribal groups in the Nunavut territory, including the Ahirmiut, Sadlermiut, Igloolimiut, and Netsilingmiut just to mention a few by name (Légaré 2002). In the 1960s, there was a push among the people in the Nunavut territory to bring about a more collective Inuit identity. Crucial factors in establishing this identity included the establishment of the Nunavut territory itself, promotion of the term "Inuit" to describe anyone actively identifying with a local tribe, and the education system throughout Canada's Eastern Arctic region (Légaré 2002).

The establishment of the Nunavut territory and creation of collective identity in the region were intertwined. After the emergence of Quebecois nationalism in the 1960s, the Canadian government understood that a collective Canadian Inuit political identity also demanded recognition. Negotiations between Inuit politicians and the Canadian federal government lasted about 30 years and emphasized the idea of Inuit nations within the Canadian state. This highlighted Inuit group uniqueness against a Canadian framework, while still recognizing their place as citizens. In 1999, the Nunavut territory was established to serve as both an Inuit homeland and a public government that fit neatly within the Canadian polity (Wilson 2005).

Though 85% of the population within the territory was Inuit when it was officially established, residents lacked a large-scale collective identity (Wilson 2005). More urban spaces allowed for more greater interaction between urban and rural residents, including all of the disconnected Inuit communities. Identification with other Inuit tribes was prompted by the distinct and numerous differences between tribal and urban life that confirmed inter-tribal similarities (Searles 2008; Searles 2010). Defining and understanding non-tribal identity shaped the Inuit identity in Nunavut.

The Inuit identity was built on the unique and unforgiving landscape, characteristic of Canada's Eastern Arctic region (Searles 2008; Searles 2010). Inuit forms of survival such as hunting, fishing, trapping, and camping in the vast arctic desert of Nunavut is a rite of passage for membership within the Inuit ethnic group (Searles 2008; Searles 2010). Belonging to the Inuit group is related to the adherence of the Inuit lifestyle.

Practicing similar lifestyles caused traditions and symbols to mold together into a cohesive Nunavut identity. Inukshuks, for example, are stone structures Inuit hunters traditionally use as reference points in the arctic desert (Légaré 2002). Standing as a testament to their perseverance, these scattered monuments are a unifying symbol, so much so that a depiction of an Inukshuk is found on Nunavut's coat of arms and flag. The development of the Nunavut identity around these traditional symbols exemplifies the deep connection to the landscape in their Indigenous lifestyle. Nunavut identity is also an example of an identity shaped by the development and understanding of an opposite identity.

Classic Regional Separatism

Most modern states arose from some sort of separatism, whether from fiery revolutions, or as a consequence of major geopolitical changes, or by mutual agreement between the state and region. Between 1900 to today, the number grew from about 53 to some 195 independent states. Figure 7.2 below shows the change in the overall number of states since the early 19th century. Whereas the 19th century was a period of consolidation, through national consolidation and empire building, the last 80 years have witnessed several trends leading to many more independent states. Much of this proliferation was a result of decolonization, as overseas and land-based empires alike divested themselves of colonial holdings either through force or because they did not have the wherewithal to administer so much diversity within far-flung territories. Other cases occurred due to the breakup of a few larger multicultural states, especially the Soviet Union, and some successful secessions, like Eritrea, South Sudan, and Bangladesh.

Even following this explosion in new countries, the remaining states still contain large numbers of regions where strong national identities are maintained. These may be considered "nascent" national separatist movements, and some of the most historic examples can be found in Europe where national-regional loyalties are found within about every single country (see Williams 1982). Beyond Europe, potential breakaway regions are found in all corners of the world. Many of these have been provided a great degree of leeway, whereas others have been brutally suppressed. In Chapter 6 we discussed how the political structure can accommodate this cultural and geographical diversity. In this chapter, we look at this from the perspective of the region and potential regional/national movement itself.

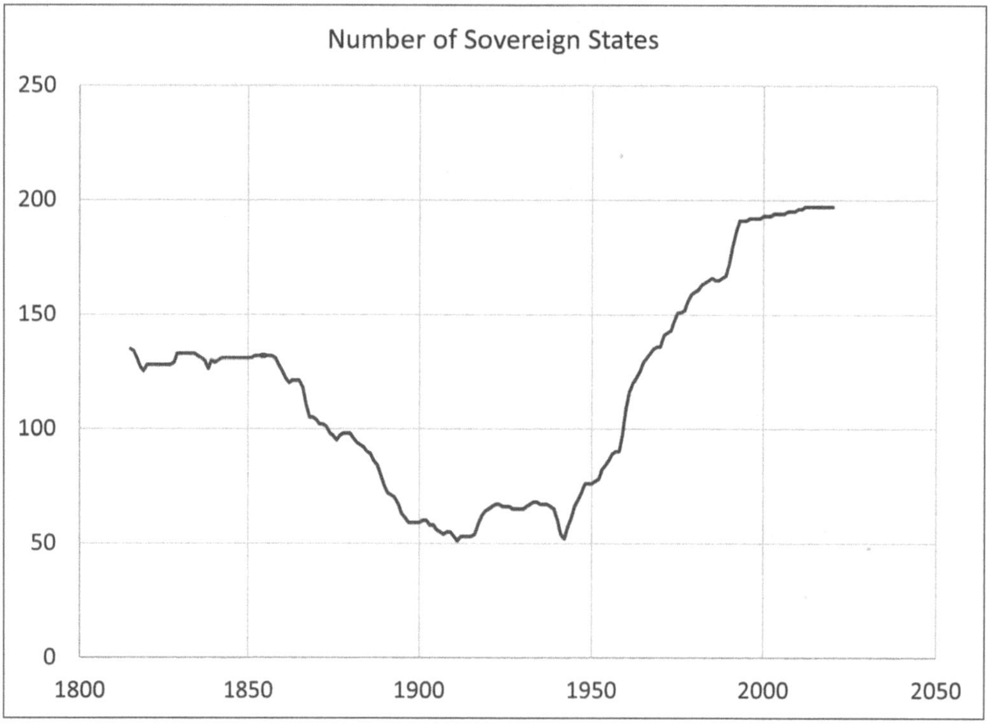

Figure 7.2 Change in the number of sovereign states
Source: David Kaplan, based on data from Griffiths and Butcher 2013

Aspects of a Separatist Movement

There are key geographical aspects to the viability of these separatist movements. The first considers the attributes of the region itself. Having a distinct cultural group is essential, of course. There must be some markers of cultural distinctiveness. Language and religion are the most common, though other markers are available as well. There must also be a sense among the members of the cultural group that they consider themselves to be a nation (Orridge and Williams 1982). Members of an ethnic group may feel a keen sense of distinctiveness – special foods, some ceremonies, a cultural legacy, for instance – but not see themselves as part of a national group.

Yet feeling a part of a nation is not sufficient, as we can see when looking at broadly dispersed populations. The critical issue is how well the minority population adheres to a territory. They may be more broadly dispersed or have little affiliation or control over the area in which they are most likely to be found. Williams and Smith (1983) laid out how this "national construction of social space" could occur and how nationalists within a region promote a sense of regional/national identity – fusing the national members to particular spaces.

The strength of national identity is not the same thing as political mobilization. Members of a region can feel very distinct but may be satisfied with current arrangements. On the other hand, sufficient reason for grievance – whether economic or political – may amplify the distinctiveness felt by members of a region. In his classic book, Michael Hechter (1977) spoke of a "cultural division of labor" among the Celtic peripheries to the British state. Members of these groups – Welsh and Scottish particularly – felt reduced to economic stations subordinate to the economic position of the dominant English. Grievances stoked a sense that separatism might be an attractive option. A feeling of being culturally stifled could also see greater political autonomy as the solution (see Smith 2009).

There is tremendous power in whether and how minority nations are territorially recognized by the state. The geographies of these nations may be demarcated in distinct provinces within each state. A country may contain a province that is marked off as the home of nation A. Or a minority nation may not be recognized in any administrative unit. This may be because the state has chosen not to recognize that nationality or simply that they do not wish to consider it as a territorial as well as a national group.

Members of a region, even within the most distinct national region, are hardly monolithic. The nation-region itself often contains members of the state majority, sometimes in quite powerful positions, along with other minority groups who exist outside the regional majority and the state majority. These can be other native groups of longstanding provenances (including Indigenous groups) or immigrant/ethnic populations that trace their lineage from elsewhere. Even among members of the national-regional population, there will be many political differences in what they see as the desired outcome and the best way to achieve these goals. Regional nationalists are often divided between more moderate parties that advocate negotiation and some measure of autonomy and fiery factions committed to whatever means necessary to achieve independence.

In Chapter 6 we discussed ways in which the state can be configured to allow for greater diversity. These political structures are important and may reflect the attitudes of the leadership. Careful examination of foundational documents, especially the constitution, can address some of the following questions. Does the central government recognize the members of the cultural group as worthy of consideration in the first place or does it deride them as nonentities – such as the manner in which the Turkish government dismissed the Kurdish people as "mountain Turks"? Does the government consider the region as a distinct territory, or does it just recognize minority individuals? Are central government officials willing to collaborate with the officials in the minority region or do they just impose their own solutions to regional distinctiveness? Fundamentally, is the government repressive or open to national difference and some measure of autonomy? These are a helpful beginning but, then again, they may serve as a smokescreen. Many constitutions devoted to cultural expression and regional autonomy are never followed in practice.

One significant item to point out: the situation of minority nationals outside their region of dominance. Members of the minority nation living outside the borders of the minority region are impacted by negotiations between minority nationalists and the state. Rights accorded to a culturally distinct population must be separated from rights accorded to a culturally distinct region. This distinction does not matter too much if special rights are provided to all members of the minority nation, regardless of where they live. But most minority nationalists prefer that their autonomism be granted within a territory; otherwise, it is more difficult to implement. Special regions do not generally extend rights to co-ethnics outside the region as these would be covered under the rules of the larger state or under the regions where they reside.

Most separatist movements are not able to prevail so easily. The balance of costs and benefits may not tend to one side or another. The overarching state may be weaker or more accommodating, but not so much as to allow a strong movement. The people and the region in question may feel a sense of national unity, but there may be enough fissures and doubts to derail the needed momentum. These movements may result in quite distinct outcomes, ranging from even tighter repression to greater accommodation. While the benefits are clear, the costs may be a great deal higher than expected. In many cases, the quest for separation resulted in bloody wars – sometimes leading to successful conclusions and sometimes not.

Where the Costs of Independence Are Too High

At times, the costs of independence or even greater autonomy are too high for prospective nations. This can be seen with three disparate examples – one where costs are greater internally and the other two where the overarching state ensures that costs will be insuperable. The case of Puerto Rico is instructive in how the internal costs of independence have made it difficult for islanders to reach a consensus. Puerto Rico has been a part of the United States since the Spanish American War ended in 1898. It is not a state and has entered into the United States under a "commonwealth status" since 1952 (Issacharoff et al. 2019). This status provides Puerto Rico with many benefits such as US citizenship, free crossing of people, goods, and services into the US mainland. Puerto Rico cannot conduct its own military policy or engage in foreign affairs. However, it is fiscally separate from the US. Puerto Ricans pay no federal income taxes, but they do enjoy federal benefits such as social security and food stamps. They also receive relief after disasters, most recently Hurricane Maria, though President Trump's slow response showed how much of a political football disaster response could become.

The status of Puerto Rico has long been ambiguous, with many aspects of the relationship unresolved. For some, the island's relative poverty benefits from an American association. The US provides security, and it has attracted a great deal of industry to Puerto Rico. Yet Puerto Rico also finds itself hampered by rules over which it has little control. Advocates for the status quo argue that severing this would have seriously harmful effects. In fact, many islanders believe that only statehood will maximize their economic potential. Yet there is also long-term sentiment for greater autonomy or independence. Depending on how it is phrased, a significant share of the vote in recent plebiscites argue for more autonomy (Pantojas-Garcia 2013). From this perspective, the economic relationship with the US has been colonial, with far too many restrictions placed on Puerto Rico's economic interests and without the full political autonomy to forge its own growth path (Berríos Martinez 1977 provides the rationale). Gautier (2018) argues that Puerto Rico needs control over policies that encourage more international trade as opposed to a continued reliance on aid from the US federal government. Yet most Puerto Ricans do not yet see the value of full independence, as measured by public opinion polls, with a slight plurality favoring statehood (Puerto Rico Report 2020).

Even given the prospect of a united region, the potential nation needs to consider what resistance will arise from the central state. States are usually quite jealous of their territorial integrity and will do whatever is necessary to prevent breakaways. Bartkus (1999) relays the case of the Nagas in far Northeastern India. Just as India was preparing for independence in 1947, so the Nagas were petitioning for their own independent country. They established a governance structure with British approval but found that the new Indian government was not willing to countenance this. As Nehru said, "No state, big or small, in India will be allowed to remain independent" (in Bartkus 1999, 58). The near unanimous support among the Nagas for independence was ignored by the Indian government. Eventually, India chose to deploy the army to quell the independence movement, jailing the leadership and forcibly moving the rural residents into walled villages (Bartkus 1999).

China has likewise placed enormous constraints on any potential for independence or even greater autonomy. Constitutional and legal documents have set out autonomous regions in China, notably Tibet and Xingjian. Yet the Communist government has long been wary of providing too much leeway and cracks down hard on even the slightest ripple of nationalist sentiment. In Xingjian, the Chinese government curtailed the authority of the provincial government to do anything meaningful, ensuring that the primary responsibilities were still within central government hands. They then reduced the power of the Uighur population (some 80% of the province) by manipulating territory, designating various counties for Xingjian's other minorities. Every minority was entitled to their own piece of territory, something Côté (2012) describes as *matryoshka* autonomy, after the Russian nesting dolls. The Chinese government then brought in millions of China's dominant Han migrants, seriously diluting the dominance of the Uighurs in Xinjiang (Côté 2012). This was accompanied by a great deal of politicking and military action against what China claimed to fear were the separatist tendencies of the Uighurs and the prospect that they might join up with other radical Muslim movements in the area (Davis 2008). To Chinese government eyes, Uighur demonstrations and some violence only lent credence to these actions and political repression further spurred Uighur resistance

Figure 7.3 Puerto Rico poster
Source: Posters of the WPA – Christopher DeNoon

(Han 2011). In the last decade, discrimination against the Uighurs has intensified. This has included greater use of surveillance, placing many Uighurs (both adults and children) in re-education camps and schools, and the destruction of Muslim religious sites (Ramzy 2021).

Box 7.2 Nationalism in Biafra

Figure 7B.2 Nationalism in Biafra
Source: Sigrid Lundberg

During the 1960s, there was a widespread demand among the Igbo people to become independent of Nigeria. In 1967, Biafra declared independence and the Nigerian government cracked down hard until the Biafran resistance fell. The Biafrans were then re-absorbed back into the broader Nigerian identity but excluded from the Nigerian political processes. Nigerian political systems, built on an ethnic basis, only enhanced ethno-nationalist rhetoric among the Igbo both before and after Biafra's days of independence (Duruji 2012; Nwangwu et al. 2020).

Though Biafra failed to achieve independence, the introduction of civil democracy within Nigeria created a space where people could re-envision an independent Biafran republic (Onuoha 2012). Collective history as the lost tribe of Israel established legitimacy as an ethnic group and provided a basis for unification (Duruji 2012). Consequently, the beginning of the 21st century has seen a resurgence of Igbo nationalism, motivated by injustices against Igbo people as well as their under-representation in Nigeria's government (Nwangwu et al. 2020). Igbo nationalism may also be seen as a way for Igbo elites to form their own power systems supported by strong ethnic identification (Duruji 2012).

Igbo nationalism is a movement against the Nigerian state, but inter-group differences based on distinctions of class and ethnicity prohibit efficient achievement of any large goals (Nwangwu et al. 2020). Espousing a non-violent philosophy, the Movement for the Actualization of the Sovereign State of Biafra (MASSOB) has amassed quite a bit of support. MASSOB uses symbols and collective Biafran memory to unify Igbo people beneath its cause and legitimizes Biafra by establishing social, political, and economic systems outside of and directly opposed to those of Nigeria (Duruji 2012). Other groups are impatient with MASSOB and seek other means of gaining self-determination (Onuoha 2013).

In terms of territorial legitimacy, the neo-Biafran movement legitimizes itself through organized and community-based action (Onuoha 2013). Individualized representation of national symbols in the form of clothing and signs, the Biafran pound, flag, passport, and anthem are highly influential (Duruji 2012). By establishing Biafran systems and rule of law, the movement de-legitimizes the Nigerian state within the Igbo homeland, using the sovereign state system against it (Onuoha 2013).

How Shifting Geopolitical Considerations Change the Balance

If the relative costs of secession are low and the benefits higher, then secession movements can be quite successful. Such was the case for some Eastern European regions in the 1990s. There were three great geopolitical events – all related to the end of the Cold War – that occurred during this period which effectively redrew the world map. The first was the rapid collapse of Communist regimes in Eastern Europe in 1989, as each of the Warsaw Pact countries threw off their dictatorships. The second was the dissolution of the Soviet Union itself, in 1991, as the huge multinational empire was broken into its 15 constituent republics – from the still enormous and multi-ethnic Russia to little slivers of country such as Moldova. The third was the dissolution of Yugoslavia as the ruler of the country since World War II, Josip Broz Tito, who had forged a path away from the Soviet Union, died in 1980. The federation held together for another ten years or so until broken apart by a desire of each nationality for greater autonomy and some form of independence. In many (though not all) of these cases, the costs of separatism went down sharply as there was no longer a powerful state to repress these political activities. This also shows how a clear consensus at the regional level can also hasten a successful break up.

Examinations of what makes successful secessions demonstrate that separatist movements within existing administrative districts – particularly first order regions such as provinces, republics, or states – are much more likely to succeed (Griffiths 2016). Existing bureaucratic structures within a first order region can form the basis of a nascent state apparatus. Minority national imaginations conceive of this district as a territorially integrated and nationally infused unit.

One successful example of peaceful secession occurred in 1993, when the country of Czechoslovakia separated into the Czech Republic and Slovakia. The process occurred so smoothly that it was dubbed the "Velvet Divorce." There were specific circumstances associated with this secession, the main one being that Czechoslovakia had just emerged from political oppression as a one-party Communist state in 1989. Before that period of time "Czechoslovakia" had existed as a unified country only since World War I and the breakup of the Austro-Hungarian Empire. Two other factors that made this transition particularly smooth were that the two populations were cleanly separated. Plus, there had been a history of good relations and elite accommodation between the two nationalities. This allowed a separation all to occur without rancor or bloodshed as both sides saw that the benefits exceeded the costs (Valková 2014).

While not as amicable, the smooth secession of Slovenia was a product of an existing administrative state apparatus, within one of Yugoslavia's constituent republics, and a strong concentration of ethnic Slovenes within the Slovenian territory. Once Yugoslavia no longer remained tenable, leaders still tried to stall any sort of secession. The two largest groups within Yugoslavia – Serbia and Croatia – engaged in a major civil war, bringing the fight particularly to the republic of Bosnia-Herzegovina and its diverse population. Slovenia, however, was largely spared this bloodshed and was able to separate out from the rest. In a plebiscite, Slovenians strongly voted for independence in 1990 and became independent in 1991 with a few small skirmishes. Of all the Yugoslav republics, Slovenia was the most homogeneous and had a rare degree of consensus (Pesic 1996). What is more, from the late 1980s, the Slovenian media, especially public television, continued to bolster the idea of a distinct Slovenian people within Slovenian territory, distinct from other Yugoslavs (Volcic 2005). This worked well, and created a fairly peaceful, prosperous, and Westernized country in short order.

How Changing Sense of National Identity Affects the Prospects for Separation

The independence of Bangladesh illustrates a case where a sense of national identity shifted enough to create incentives for two separations. There was no major geopolitical shift that encouraged the independence of what was then called East Pakistan, just a change in the way East Pakistanis viewed themselves. Prior to Indian and Pakistani independence, Muslim India did not necessarily consider itself in primarily territorial terms. The notion among the early leaders, such as Jinnah, was that a united India would create a system guaranteeing its Muslim population a voice. Yet there was also some notion of territorial separation given the existing population distributions. Many Muslims were found throughout northern India but were especially concentrated in the northwest and northeast sections. Long before Indian independence, there had been several proposals regarding the division of British India between its Hindu and Muslim majority regions. In fact, Pakistan was an acronym made up of several Muslim provinces in India's northwest (Das et al. 2020). Even as partition proceeded, with the failure of any consortial agreement, many Muslims (and Hindus) expected to remain in their own communities (Harper 2010). Yet this did not occur as the partition of India and Pakistan in 1947 led to a huge population transfer between the two countries, as a steady stream of Muslims and Hindus found themselves in a now-hostile territory and sought to move to safety.

One other consequence was the creation of Pakistan as a country with two distinct parts on either side of India. When partition was eventually granted, the Bengali-speaking regions in the northeast were added to the new Muslim country. This region was different in several ways. There were fewer migrants moving to East Pakistan as compared to West Pakistan. East Pakistan was also demographically larger, about 20% larger (Siddiqui 2012).

The creation of East Pakistan after partition was further beset by two issues. In this area, the Bengal language region stretched beyond its Muslim-dominated region. Bengal had been separated between Muslim and Hindu adherents at the turn of the 20th century with predominantly Hindu West Bengal, including the Indian city of Kolkata, and predominantly Muslim East Bengal (Das et al. 2020). But there was no clean split between the two areas, with adherents interspersed to an extent that, even today, the boundary between India and Bangladesh includes scores of exclaves and enclaves (Jones 2009). Hindu Bengali did not recognize their Muslim co-linguists, referring to them instead as simply "Muslims." Strong economic disparities overlaid this religious division, with the Muslim population underrepresented and impoverished, motivating separatist sentiments. Within the Muslim community, the second issue lay in the unclear status of language and religion. The small numbers of urban, upper-class Bengali Muslims preferred Urdu, relegating Bengali to a peasant-based language without much standing and sparking a great deal of resentment among the Bengali-speaking lower classes. Unified Pakistan built on this sentiment and denied Bengali any place as an official language, promoting Urdu instead. Despite East Pakistan's larger population, West Pakistan exploited East Pakistan economically and marginalized it politically. This echoed the economic disparities that had earlier been suffered by the Muslim Bengalis vis à vis the Hindus (Kabir 1987).

The incentives for East Pakistan to gain greater autonomy cascaded to the point where, by the 1960s, many residents of East Pakistan considered the situation untenable. One key factor, though, was a regional identity-shift of the region from an emphasis on religion – used to distinguish themselves from the neighboring Hindu Bengalis – to language (Kabir 1987). This linguistic nationalism was expressed in a 1966 manifesto demanding that massive discrepancies between East and West Pakistan be eliminated and that East Pakistan be conferred significant autonomy. The response by Pakistan's military government to these demands was to arrest the authors, beginning a four-year war of secession, massive hardship in East Pakistan, and a brutal campaign against all Bangladeshis suspected of being part of the liberation movement (Das et al. 2020). Bangladesh eventually became independent but continues to experience the tension between the religious-Islamic and the linguistic-Bengali aspects of its identity.

Autonomy within Special Regions

The identification of a national identity along with the mobilization of members of that nation for greater political autonomy need not always lead to independent states. There can be a middle position between complete sovereignty and zero political recognition. How such regions are managed at the level of the state is discussed more fully in Chapter 6, but often entails different degrees of federalism and devolution. Federalism is a mechanism whereby the central and regional governments each enjoy sovereignty within their respective spheres of influence – with a great deal of variation in the relative power of regions. Devolution is an example where the central government, often but not always within a federation, transfers a set of powers to a regional government. Regions may enjoy control over vast aspects of life – including legal systems, languages, education policy, and immigration. It is within this context that many countries contain so-called special regions (whether designated by that title or not). These regions are provided with additional autonomy, allowing for national identities to be preserved but within a state structure.

The case of Quebec demonstrates how this can be accomplished more or less successfully. Quebec has always been the Canadian province containing most (but not all) of Canada's French-speaking population. It also includes a large English-speaking population (smaller than has been the case traditionally, but still quite substantial), a high volume of immigrants and their descendants, and many Indigenous peoples found in various places throughout the province. The French population in Canada has enjoyed a special status as one of the two founding groups (along with the English) of Canada. The binational character of Canada has been apparent at the federal level since Canada's independence in 1867. French-Canadians have been instrumental in running the country – about half of all prime ministers have been French-Canadian – and Canada itself is an officially bilingual country.

Relations between the French-Canadian population and the Canadian state, as well as the English-language majority, have evolved over time. French Canada has long concentrated in Quebec but has spilled over into many adjacent provinces such as New Brunswick and Ontario, as well as in small pockets across the entire continent (Kaplan 1994a). (There is also a significant diasporic population across the international border in New England.) From the late 19th century to the mid-20th century, French-Canadians were in many cases relegated to second-class citizenship. The economy, even in Quebec, was conducted in English and the onus was on the French-speaking population to learn English in order to get ahead. This was even true in Quebec, where businesses, major retail concerns, and the bourgeoisie was all an English-speaking affair. Much of the French-Canadian population was agricultural and quite insular (Kaplan 1989).

Much of this changed after the 1960s. The French-Canadian population as a whole became far more progressive, and they began to chafe at the clear inequities. Many partisans cried out for greater control of their own affairs. They also began to harness the fate of French Canada increasingly to the province of Quebec itself, proposing a cultural-territorial solution (Kaplan 1989). New political leaders also emerged, and a new political party, the Parti Quebecois, was formed around an explicitly separatist objective.

Grievances against the dominant Anglophone population turned violent in the late 1960s. Montreal was racked by terrorist activities of the Front deliberation du Quebec. The FLQ would detonate bombs, especially in Anglophone neighborhoods of the city, and also try to blow up major institutions such as the stock exchange (Munroe 2009). This culminated in 1970 when the FLQ kidnapped Quebec's Deputy Premier and a British diplomat. Canada's Prime Minister, Pierre Trudeau, invoked emergency authorization to oversee the crisis and similar events have not surfaced.

When the Parti Quebecois gained control of the provincial legislature, it passed legislation making French the sole language of the province. The language law has had repercussions in all areas of Quebecois life. This became quite visible in the discursive landscape where French-language signs are mandated throughout under penalty of strict fines. Since there remains a large English-speaking population, one can walk around a neighborhood where the language on the street is English, but the signage is all in French.

The second major change occurred in education, where the language discrepancy was addressed by requiring immigrants (including Anglophone migrants from other provinces) to attend French-language schools. This effectively moved the linguistic education boundaries but retained the strong split in schooling. Likewise, the assortment of Anglophones into particular towns has provided them with some measure of local political power. Even Quebec's language law, which mandates all signs be only in French, covers the strong differences in language use among the individual neighborhoods. Yet the separation between the two linguistic groups persists, and it also pans out in their separate spatial identities, as Anglophones look to the country of Canada, whereas the Francophone population is invested primarily in Quebec (Kaplan 1994b).

Quebec itself includes a spectrum of political thought, from die-hard secessionists to those more interested in accommodation. Since French Canada's affirmation of its more nationalist determination, the politics of secession have proven tricky and plebiscites in 1980 and 1995 turned down the separatist option. In the 21st century, independence has become less popular still. The latest opinion poll showed it losing by 56% to 32% (Fournier 2021). And in 2018, a new party prevailed that was against independence but for greater provincial powers (Montpetit 2018).

Finally, Quebec contains a large number of linguistic and ethnic minorities. Within this group are included members of the English-language population who live primarily in Montreal but can be found throughout the province. In the latter 20th century, English speakers lost what had been a most privileged standing, with control of the economy and the ability to live their lives entirely in their mother tongue. Language laws enacted by Quebec's government made their standing less comfortable, leading in part to widespread exodus to other Canadian provinces (Kaplan 1994a). Other immigrant groups arriving in Quebec were also mostly interested in acculturating into the English language, which has resulted in Quebec's government seeking a much stronger hand in immigration, with a preference for French-language immigrants. Finally, the native peoples in Quebec have sought their own autonomy (Scott 2011).

Nationalism within Dispersed Geographies

Nations that are completely enclosed within a single state threaten instability. Nations that straddle state borders complicate the situation even further. It has influence whether a group straddles two countries or whether it ranges over multiple countries. In the first instance, there may be a tendency towards irredentism, where the national group in one state simply seeks to join an adjoining state, effectively shifting the boundaries. (Such a claim can be supercharged if the adjoining state also claims that territory.) The second instance, where a national group straddles multiple states, brings multiple actors into the geopolitical equation.

Box 7.3 Ukrainian Breakaway Regions

Figure 7B.3 Ukrainian breakaway regions – pro-Russian demonstration in Donetsk 2014
Source: Andrew Butko, CC BY-SA 3.0, https://creativecommons.org/licenses/by-sa/3.0, via Wikimedia Commons

In the spring of 2022, after a Russian declaration of war against Ukraine, Russian president, Vladimir Putin, announced Russia's recognition of the separatist Donetsk People's Republic and Luhansk People's Republic. Both "Republics" are part of the Donbas region of Ukraine and are home to large Russian-speaking populations sent there by the Soviet Union after World War II to work (Aljazeera 2022; Mirovalev 2022). Donetsk and Luhansk were originally founded by English industrialists John Hughes in 1869 and Charles Gascoigne in 1795, respectively (Mirovalev 2022). Both share Russian and Soviet history, being part of the Eastern portion of Ukraine that was previously annexed by czarist Russia and part of the Soviet Union along with the rest of Ukraine.

In 2014, Ukrainian President Viktor Yanukovych was removed from leadership after he had refused to agree to greater trade cooperation with the European Union. A government more aligned with Western European progress was then instated (Hernandez 2022; Westfall 2022). Russia responded by supporting those Ukrainian regions with high Russian-speaking populations (Socor 2014). This began with the annexation of Crimea which fell to Russia quite easily, and then continued in Donetsk and Luhansk where Russian-speaking populations have fought for autonomy with Russian encouragement (Aljazeera 2022).

Since 2014, both regions have operated quite differently because of their Russian-backed leadership and dependency on Russian social, economic, and political support (Kasianenko 2019; Westfall 2022). The Donbas region has been at the forefront of Russia's efforts. Russian passports have been issued to citizens of Donetsk and Luhansk and Russia has focused on establishing a nationalist narrative for the Donbas region (Westfall 2022). This uniqueness is what would make an autonomous or fully Russian Donbas region legitimate, which is why we see local support for either autonomy or being incorporated into Russia (Westfall 2022).

The insistence that Donetsk and Luhansk are effectively Russian regions has widened into the discourse about Ukraine as a whole. The area of what is now Ukraine goes back long before the development of a Russian polity. Over the centuries Ukraine passed through multiple empires and states. In the 20th century, it was a constituent member of the Union of Soviet Socialist Republics (though in an odd quirk of history, Ukraine SSR maintained a separate seat at the United Nations from 1945 to 1991). After the dissolution of the Soviet Union in 1991, Ukraine became independent and its relations with Russia were fairly civil. But with the regime change in 2014, Putin became more adamant that all of Ukraine was a Russian region and is now trying to enforce this vision with military action (Westfall 2022).

Classic Irredentism

Cases of irredentism occur when a regional-national group in one state seeks to join up with their co-nationals in another state. Irredentist movements are found throughout the world and often result from wars where territory is offered to a country or by partitions that are not as clean as they might be. Relations between the states involved also figure into the politics and identity of irredentist groups, emboldening or weakening them depending on whether the adjoining state has a claim. There have been a few successful irredentist movements in recent years (Crimea is among the most recent) but most of them are likely to fail in the absence of major geopolitical shifts. Greater satisfaction with the status quo among residents of a region can also play a role. At the same time, there are many cases where irredentist sentiment runs deep throughout the population.

Since irredentism is an Italian word, it makes sense to consider an irredentist movement within Italy. The South Tyrolean case illustrates this classic irredentism (see Figure 7.4). It is an area where speakers of German predominate and who feel a strong kinship with the Austrians across the international border, and specifically Austria's Tyrolean region. The vast majority of such German speakers live within the Italian Province of Alto Adige, and this is also recognized administratively, though German residents refer to this as the SudTirol.

Up until World War I, this province belonged to Austria, along with Trentino to the south. It was then taken over by Italy and remained a part of Italy for over 100 years. This land was once part of the Austrian Tyrolean province but was acquired by Italy after World War I. After World War I, Italy was able to expand its land area, gaining the southern Tyrolean region and the Istrian Peninsula from Austria-Hungary. At the same time, Benito Mussolini was consolidating power. Mussolini, as most fascists do, sought to erase the German language element from the newly acquired region through an Italianization policy. In the South Tyrol for example, Mussolini forced German speakers to

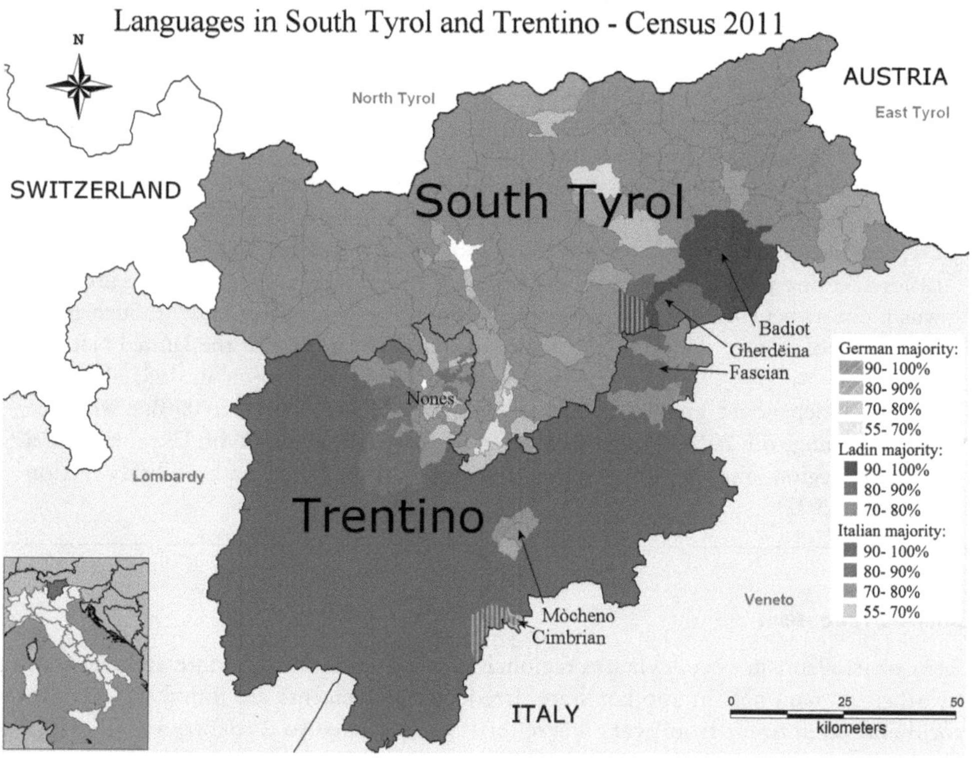

Figure 7.4 Case of the South Tyrol in Italy
Source: Stephanie Burkey

change their names, he converted all place names to Italian, he prohibited speaking German, and he encouraged a widespread population influx from southern Italy to the region, pushing many German speakers out into the countryside (from Kaplan 2000). Even the name of the region itself was changed – to Alto Adige – and individual cities like Bozen were renamed Bolzano.

The degree of rancor between the German-speaking peoples in the Alto Adige and the Italian state has calmed. During the inter-war period, Benito Mussolini squashed any cultural expression and tried to uproot all elements of Austrian heritage from the people and the landscape. After World War II, the newly democratic Italian government reversed Mussolini's cultural bans and granted the region some level of autonomy, though the effects of Mussolini's policies, and in particular the installment of many Italian speakers in the region's major cities, lingered. Nationalists sought to return the region to Austria, leading to some violent conflicts in the 1950s and 1960s. Austria continued to exert influence, especially inasmuch as it first claimed the South Tyrol, and even after abandoning these claims, continued to see itself as a protector. The level of rancor led to arbitration by the United Nations in the early 1960s (Alcock 1992).

Sudtirolean irredentism has tapered down in the last 50 years. Austria and Italy declared their dispute at an end in 1969 and the major geopolitical change was in the strengthening of the European Union and the opening up of international borders between Austria and Italy (Alcock 1992). Practical considerations have also tamped irredentist sentiment as the province is among the wealthiest in Italy and it receives a large number of tourists from both Austria and Germany (Kaplan 2002).

From the standpoint of German residents, the Alto Adige province has been provided with a great deal of autonomy, including status as a "special" region and the development of a strong German-dominated regional party, the Südtiroler Volkspartei, which fights for strong German language rights. The landscape itself seems to mostly reflect a German orientation as beer stubs and gasthofs are common features. Yet this discursive landscape is still contested. One of the biggest political issues is whether the signs should continue to have both German and Italian names (as agreed earlier) or only German names. Some people decide to cross out the Italian names on these markers as a way to emphasize their irritation at any Italian identity claims here.

Northern Ireland shows how such sentiment can transform over time as several underlying circumstances change. Ireland was long possessed and exploited by England. Ireland had been mostly conquered by England during the Anglo-Norman invasions of the 12th century. While England and Scotland shifted to Protestantism in the 16th century, the Irish population remained Catholic. Later, in the 1600s, Ireland was intentionally colonized by English and Scots in a series of "plantations" – creating pockets of Protestantism – most far wealthier than the Irish Catholic population. After decades of strife, Ireland was finally given autonomy in 1922 and full independence 12 years later (Stamp 2014), but the northeastern counties – which had the largest Scots-Irish population – remained as part of the United Kingdom.

Since this time, the population in Northern Ireland has been strongly divided between Protestants who wish to remain in the United Kingdom and most Catholics who would dearly love for this territory to rejoin Ireland. Tensions between the two groups heated up in 1969 when the "Troubles" brought widespread sectarian fighting throughout the entire province (Figure 7.5). The Republic of Ireland continued its claims on the north while Britain sent its military to keep the peace but widely seen as a partisan combatant.

Figure 7.5 Belfast memorial mural to the Troubles in the summer of 1969
Source: Albert Bridge

In addition, the Protestants were worried about being demographically outnumbered as their initial majority gave way to rough parity with Catholics (Kaufmann 2011). De-escalation occurred with a change in the nature of some of the nationalist parties like the Irish Republican Army and the signing of the Good Friday Agreement which provided the Irish Republic a say in Northern Ireland affairs. Lowering polarization, as described by Hancock (2016), has caused the "core" identities of Protestants and Catholics to fade and opened up prospects for more bridging identities.

Thanks to the Good Friday Agreement, the Republic of Ireland already has some power over the affairs of Northern Ireland which requires agreement between the two communities of Northern Ireland, between the Republic of Ireland and Northern Ireland, and between the Republic of Ireland and the United Kingdom (British-Irish Agreement Act, 1999). As populations transition and forget much of the rancor of the past, a couple of changes may make for a successful irredentist movement. First is the fact that the Republic of Ireland has become one of the wealthiest countries in the world, making unification a far more attractive prospect to more Northern Irish residents. The second is that while the United Kingdom decided to exit the European Union, Ireland remains a strong member. As a concession, the border between Northern Ireland and the south – open in the past – will continue to remain so into the future. The Brexit Withdrawal agreed to continue an open border in Ireland, which means that the real border crossing will be across the Irish Sea (Hayward 2018). If Ireland functions as a singular economy, apart from Great Britain, a lot of de facto irredentism would already be taking place.

Nations Straddling Multiple Borders

Throughout the world, there are several nations located apart from existing state configurations. These often straddle multiple states and associated boundaries. This has long been an historical fact. In the past, numerous nations were fragmented across states and kingdoms – Poland being one of the best-known examples. Other nations stretched across numerous smaller units as a legacy of Medieval times – sometimes remedied by the creation of whole new countries such as Italy and Germany in the 19th century.

More states mean more actors involved, each with a different sort of relation to the dispersed nation. Members of the nation may be recognized and accommodated in one state while find themselves shut out in another state. The larger number of states also means that mobilization of the nation is that much more difficult. National sentiment and identity vary by context and the lack of easy interaction hinders the opportunities for a singular political project. Members of the same nation may hold dear their identity in one context, acculturate to their neighbors in another context, or develop a hybrid identity in yet a third (Laitin 1998, 32–34). Politically, the nation can find itself in many different circumstances. They may have a semblance of a political homeland in one area – e.g., the Palestinian Authority – but remain well dispersed beyond this.

Kurdistan has been called the largest nation in the world without an independent state. The Kurdish people straddle at least four contemporary states: Iraq, Turkey, Iran, and Syria and have occupied this land for well over 1,000 years (see Figure 7.6). They are Muslim, like their neighbors, but are otherwise culturally and linguistically different from Arabs, Turks, or Iranians. Like so many proto-nations, prior to the 19th century, most Kurds did not have a sense of an overarching identity; they were comprised of localized communities distinguished by their language. The ethnic term "Kurd" was likely acquired in the 6th century BC and the Kurds themselves adopted Islam in the 7th century AD while retaining all other aspects of cultural distinctiveness. As members of the Ottoman Empire, most Kurds were allowed to retain some cultural autonomy under small feudal provinces (Culcasi 2010).

Ethnic Kurdistan, a large vaguely formed region primarily in Ottoman lands but also in parts of the Persian Empire, developed national sensibilities towards the end of the 19th century. Many credit Sheikh Ubaydallah for unifying the Kurdish tribes and for instilling a sense of national consciousness. Ubaydallah wrote that "The Kurdish nation is a people apart … We want our affairs to be in our own hands" (in Olson 1991). The two principles of nationalism were expressed here in that the Kurds were culturally distinct and also bound to determine their own fate. With the changing geopolitical situation following World War I, it appeared that Ubaydallah's dream would be realized. The 1920 Treaty of Sevres was signed allowing the formation of a Kurdish homeland at least within what was to become Turkey. But Turkish nationalism and the hostility of Kurds living in other lands put an end to this idea and since then Kurdistan has remained a dream fractured by Kurdish location in multiple countries and varying

Figure 7.6 Map of Kurdistan
Source: Stephanie Burkey

political treatment of Kurds. In Turkey, differences in culture and some distinct physical features have caused the group to be racialized; many Kurds have been dismissed and persecuted (Ergin 2014).

While Kurdish ambitions have been quashed in Turkey and Iran, in the last 30 years a quasi-independent Kurdistan has emerged from the demise of Saddam Hussein's Iraq (Jüde 2017). This began in the aftermath of the first Gulf War when a "no fly zone" was imposed on the northern area of Iraq where the Kurds are concentrated. The freedom from Iraqi control allowed Kurds to develop the functions of a state, creating an elected parliament, welfare provisions, and a strong military (though hampered initially by fighting between two rival organizations). This Iraqi Kurdish quasi-state persisted after Hussein's reign ended and is likely to remain a fixture into the future.

Whether this Kurdish quasi-state will indeed become the world's newest country is still unclear. There are recognition issues from regional powers as well as the United States (Caryl 2015). What is more, the development of an effectively independent Iraqi Kurdistan only partially fulfills Kurdish nationalist demands because this territory comprises only about one-fifth of the entire Kurdish population (Jüde 2017). Ironically, the country that has been most repressive to its own Kurdish population – Turkey – is the main regional supporter of Iraqi Kurdistan. If Turkey agrees to independence, the long goal of an independent state for Kurdistan may become a reality.

Conclusion

Ethno-nationalist movements run contrary to the model of multiculturalism that exists in many strong modern democracies, yet the idea of national self-determination along ethnic lines persists as a strong motive for independence. Many ethnically distinct regions continue to vie for independence due to minority subjugation and a real or perceived sense of disenfranchisement within a system which favors an ethnic majority. Despite the goal of a homogeneous nation-state, countries become diverse through colonialism, migration, territorial expansion, creating fractures in a national identity based on ethnicity.

Granting every unique regional ethnic group independence would lead to anarchy, as political aspirations of even the most homogeneous groups are rarely unified. Rather, states and international organizations use a spectrum of concessions to appease would-be separatist groups. These concessions range from mere recognition to complete independence. Most regional independence movements are satisfied with cultural concessions and political participation, such as an allocated number of parliamentary seats, and the right to teach using a regional language. In any would-be separatist case, a cost-benefit analysis takes place both on the side of the state and of the minority region. The extent of the cultural threat, the economic prospects, and the viability of stable statehood are all either formally or informally considered.

Separatist organization tends to be much easier when the group in question occupies a contiguous territory in one state. Group territories that span multiple states, like Kurdistan, face an additional hurdle of negotiating with more than one federal government. Groups that are non-contiguous at all, like diaspora groups, must work in other ways to consolidate concessions in a transnational arena.

Chapter 8

Nationalism at the Local Scale

Nationalism is primarily considered at a "national" scale: a larger unit that comprises considerable territory and both rural and urban terrain. However, it is also a vital aspect of the local scale. Cities assume an outsized meaning to the nation. It is often where the ruling class lives and governs. Historically, most organized states began in the cities. As certain cities prospered, they acquired more territory – a hinterland – from which they could command resources and people. Sometimes they would grow to incorporate other cities, or to establish new cities in their domains.

The city has maintained its importance in the era of nationalism. The capital city and/or the largest city – sometimes but not always one and the same – represent the country and often the national majority. They are the last redoubt. If the city falls, that signifies the final defeat of the country, whether it be the 1975 fall of Saigon after the Vietnam War or the 2021 breaching of Kabul by the Afghan Taliban. The city represents something different from the country in terms of its "national" composition. In most countries, the big cities are where the nation bends a little. While rural areas may be fairly homogeneous, the city is often quite diverse. The people living there are also more likely to have different attitudes. Ironically, while the main cities are the seat of a country's power, often nationalists (and especially ethnonationalists) define the "real" nation as residing in the small towns and rural reaches outside of the city.

This chapter will address some of the most significant aspects of how this operates on the ground and in the national consciousness. But first, it is important to make a separation between the role of the local area as a generic sort of territory, and of certain local areas – such as the capital, largest city, symbolic hearths – as crucial to the national identity. Brussels, Belgium looms large in both Flemish and Wallonian nationalism as a capital and center; Jerusalem is both an idea but a vital place in the imaginations of the Israelis and Palestinians. Second, we can consider just how significant the urban landscape is to national identity. The national landscape, as discussed in Chapter 4, expresses national identity and the contestations of how national identity is framed. The local venue is where these elements are showcased. Even the tiniest village will provide ample sources of national symbolism on the landscape. Big cities and capitals amplify these symbols, as they occur more frequently and at a grander scale. Most French cities contain triumphal monuments but only Paris has the Arc de Triomphe.

National Divisions within the City

Among the abiding mysteries of nationalism is the expectation that people who are complete strangers to one another will, in crucial instances, sacrifice for one another. The

DOI: 10.4324/9780429435782-8

nation is, to many of us, an abstraction dominated by symbols of national affiliation. The abstraction becomes real when members of a nation encounter members of other nations, whether as tourists or in longer term encounters such as work or study in another country. Conflict between nations can be equally abstract for many, as fighting often occurs far away. Yet it becomes quite real when face to face with a person one has been encouraged to consider the "enemy."

In cities divided by distinct national groups, the sense of national identity, of nationalism, and of national conflict is not an abstraction but something encountered in day-to-day life. It is important here to distinguish the "ethnic" from the "national." The two terms overlap, but ethnicity is a far broader term that categorizes groups based on some cultural difference, whether objective or perceived. Like gender or age, ethnicity is one way in which individuals define themselves. Because of their greater density and economic opportunity, cities often attract many different ethnic groups – either folks coming from inside the country as part of a rural to urban or urban to urban migration, or people arriving in the city from abroad, or the descendants of these individuals who have maintained their ethnic identification. Ethnicity is a broad concept that comes in many flavors (Kaplan 2017). Yet there is something specific when ethnic differences also correspond to national differences: when members of an ethnic group desire to form their own state or achieve much greater autonomy. There is a political difference between an Igbo who has immigrated from Nigeria to the United States and a member of that same Igbo ethnic group living within Lagos.

Several cities in the world contain at least two distinct national groups. These usually include many other ethnic groups, but the national division dominates the political and social atmosphere. Some of these cities have experienced civil warfare between neighborhoods. Others have existed in a cold war of tension. Some cities have learned to live with their strong national divisions whereas others have not been so successful. In any event, the strong national divisions that exist within the country are present in the city in a specific manner. Compared to some traditional vectors shaping urban land use, such as income, geopolitical considerations determine urban morphology. Segregated spaces are developed; sometimes such spaces are demarcated by walls or other barriers (Calame and Charlesworth 2011).

Local Boundaries and National Tensions

The importance of these divisions goes beyond simple patterns, however. The relations, as expressed both socially and spatially, reflect the larger national context. Depending upon the state of relations between groups, movement across spaces may become virtually impossible. This suggests that hostility is deeply embedded within the society at large. In an influential study from half a century ago, Frederick Boal (1969) examined what he called "activity segregation" between Irish Catholics and Protestants in Belfast, Northern Ireland. Within the city during the 1960s, there were clear segregation patterns between the groups, especially pronounced in more working-class neighborhoods. Yet this was only part of the picture. The very movements of residents were dictated by lines both visible and invisible, causing people to go out of their way to avoid transgressing a rival space. Beyond this, Boal found that media consumption, sports fandom, even the naming of the neighborhoods fell along ethnic-religious-national lines. What was happening on the ground in Belfast presaged a level of tension later manifested by the onset of broader national conflicts.

The augmenting of hostilities between national groups has pronounced local effects. An extreme case occurred in Beirut during the 1980s. The Lebanese state long existed under a consortial arrangement between the Maronite Christian and Muslim populations. These two major groups occupied separate sections of Lebanon's capital and largest city, with clearly demarcated neighborhoods. For much of Lebanon's early history as an independent state, the two groups coexisted well enough such that Beirut was considered an attractive and cosmopolitan city. But in the run-up to an effective civil war in the 1980s, spatial barriers between the groups hardened, public space that had previously been shared disappeared, and moving between ethnic neighborhoods became impossible (Genberg 2003). The urban boundary between neighborhoods served as a political boundary between hostile forces.

A few miles south of Beirut, the city of Jerusalem epitomizes a city divided but one where the boundary between national groups has shifted with power and population change. As a center for three major religions – Judaism, Christianity, and Islam – Jerusalem packs major symbolism into its very existence. More practically, it has long been separated between its Jewish and Muslim residents. For several hundred years, Jerusalem was ruled by the Ottoman Empire which allowed each group to develop its own neighborhoods in line with the general Ottoman policy of recognizing separate ethnic communities and giving them some measure of self-governance. A green line was then placed under British control in the 20th century, separating the Arabs from the Jews. With the establishment of Israel, the international boundary between Jordan and the new Jewish state ran right through Jerusalem, one of the few occasions in Jerusalem's history where an international boundary corresponded to municipal divisions. But Israel's victory in the 1967 war gained it the entirety of the city. Israel's government accommodated the expanding Jewish population by shifting the green line eastward (Shlay and Rosen 2010). This has become enormously controversial among the Arab population and is seen as the sort of land grab that Israel has executed since its inception (see Figure 8.1). The relations between its religious neighborhoods have become a focal point of the Arab–Israeli crisis across the Middle East.

Establishing Parallel Societies

Well-established ethnic groups have long been able to coordinate many elements of residential life within the ethnic community. Such features as ethnic churches, festivals, welfare societies, businesses, schools, even entire ethnic economies have arisen within cities such that we often speak of "Koreatowns," "Little Italy" or "Curry Mile" to label this fusion of ethnicity, function, and place. In cities where national differences also apply, this "institutional completeness" (Breton 1964) will expand and extend to other aspects of life, most particularly the political sphere. To the extent that national groups feel themselves to be distinct, adhering to a separatist agenda, their urban communities will reflect that.

Belgium is divided into two distinct language groups, each predominating in particular territories – in this case the Flemish (Dutch) population in Flanders and the French population in Wallonia. There is also a small German-speaking community in the eastern edge of Wallonia. The capital and largest city of Brussels is also divided by the two dominant national groups, with many immigrants. It is also contained within the Flemish language region. While the largest number of residents speak French, though many are not originally from Wallonia, Brussels is considered a unique administrative entity that is

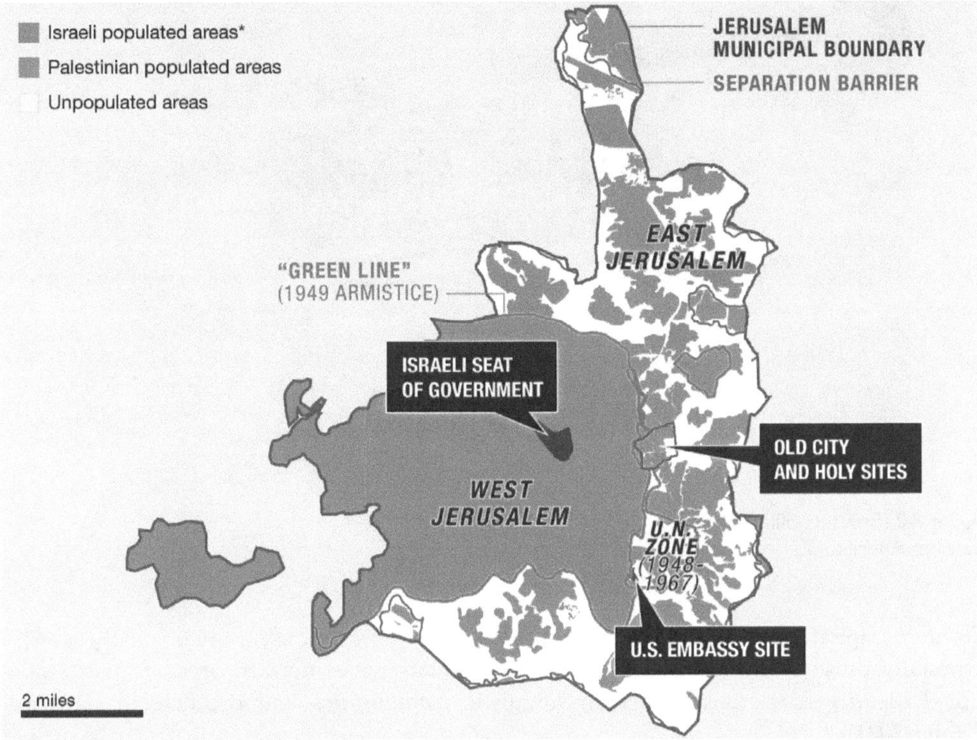

Figure 8.1 Map of division in Jerusalem
Source: Map data by Daniel Seidemann/Terrestrial Jerusalem. Labeling by NPR. Credit: Daniel Estrin, Alyson Hurt, Larry Kaplow, Brittany Mayes and Greg Myre/NPR

officially bilingual (Murphy 2002). Though a lack of linguistic data since 1961 means that we cannot demarcate separate Flemish (Dutch) and French neighborhoods within Brussels, such neighborhoods can be ascertained by other means (Murphy 2002). However, the city is not organized in such a territorial manner, in sharp contrast to Flanders and Wallonia which exclusively use Flemish or French within each territory. Instead, the capital region operates under the condition of non-territorial autonomy. Each linguistic community regulates and finances institutions that furnish cultural, educational, welfare, and healthcare services to residents. Residents are free to choose whichever service they like and how to communicate with local authorities, no matter where they reside (Mulle 2016).

Belgian residents are all quite aware that their ethno-linguistic divisions are also national divisions. Each community contains strongly nationalistic political parties and Belgium is structured to valorize these divisions and provide as much autonomy as is possible within a single state. Yet Brussels stands out both as a microcosm of the national divisions between Flemish (Dutch) and French but also as the one place in Belgium where the national divisions are tentatively bridged. Communities provide language services to anyone who wants them, and people in Brussels do cross linguistic lines. For instance, a Francophone family may decide to send their child to a Dutch school because they think it is superior (Mulle 2016). This means that Brussels is the one place in Belgium where a more bilingual sensibility might occur (Figure 8.2). The further position of Brussels as the

Figure 8.2 Brussels bilingual sign
Source: Andrijko Z.

de facto capital of the European Union also shifts perceptions, as many in Brussels – especially those working for the European Commission – may incorporate a supranational identity that transcends their linguistic communities and the Belgian state (see Calder 2021).

The Symbolic Weighting of Capital Cities

National identity includes all of the peoples who identify with the nation (and sometimes a few who do not) and the territory that national elites and common folk feel belongs to them. Yet this territory is variegated. It is composed of sites that carry deep meaning to nationalists, or of places which are essential to the national economy such as mines or ports, or of locations where major battles were fought or historic relics were uncovered. Every territory includes these sorts of places which loom large in the national imaginations. As a result, the placement and activities occurring at local sites are significant from the national standpoint as well. Within this, we can identify the capital city as a particularly salient local example of nationalism.

On one hand, the capital city is a purely functional entity. This is the place where the principal organs of state government reside. The head of the country, be it a king or a president, usually has an official residence in the capital. The legislative body meets in the capital. And many of the executive functions of the national government are based in the capital, especially the higher-level functions.

On the other hand, the capital is enormously symbolic to the state and the nation. Even colloquially, people may reference a country by the name of the capital city: "Cairo" stands in for Egypt as its seat of government. For residents of the country, the capital city represents the best and worst of their society. It can be a point of pride, a place where everybody would like to visit someday, but it also evokes all of the frustrations of government and bureaucracy. The capital signifies red tape and out of touch civil servants. The capital's role in nationalism heightens this for everybody who belongs to the dominant nation. To them, the capital represents them and their aspirations. For those who do not belong to the nation, the capital can seem like an alien force.

For these reasons, the designation and location of a capital city is quite significant to the national project. There have been several capital cities, primarily in European countries, which emerged out of a long history and reflected the dominance of a particular national group. In these cases, the capital city was also the country's most important political or economic center. Paris, located in the Île-de-France, emerged as the Capetian Monarchy stretched beyond the Île-de-France and eventually covered most of France's modern territory (Parker 1994). London was likewise based on the growing pre-eminence of Wessex in England and then England in Great Britain (Blair 2003). By the 1500s, it was the pre-eminent urban center; even more dominant than it is today (Hall 1998). The unification of states such as Germany and Italy also established their capital cities. Berlin was the capital of Prussia, which ended up consolidating the other German states in the middle 19th century. After World War II, Berlin was divided by a fortified wall running right through its neighborhoods (see Figure 8.3). With the reunification of Germany in 1989, the capital of Germany – previously divided between East Berlin (for East Germany) and Bonn (for West Germany) was again situated in Berlin. The unification of Italy did not initially include Rome, but once that city was wrested from Papal control, it was designated as the new capital of unified Italy, with the entire city being repurposed to house the state apparatus (Mazzoni 2010). For Italians, Rome had always been a capital, of an ancient empire, of Christianity – so it was only logical that it become a capital again.

Size advantages also accrued to many capitals. In the 17th and 18th centuries, capital cities hosting the monarch began to increase the bureaucracy. This growth in turn led to more overall urban growth. The "Court" became the pre-modern bureaucracy, and its

Figure 8.3 Berlin Wall
Source: David Kaplan

growth of functions also necessitated a greater reliance on public monies often extracted from the peasantry (Ringrose 1998). As governments have expanded around the world, capital cities have grown tremendously, employing larger segments of the workforce, and spawning ancillary industries of lobbyists, interest groups, and other organizations that find headquartering in the capital to be their best choice. As a result, capital cities exploded in size and importance. For many European countries, their capitals also became capitals of vast continental and overseas empires, further expanding their scope.

With only a few exceptions (e.g., Bern in Switzerland) most European countries followed the logic that the capital city was also the largest city. Yet this has not necessarily been true in other parts of the world. In many instances the designation of a capital city – or the relocation to a new capital city –provides a fresh face to the nation and epitomizes national ideals (Diener and Hagen 2018, 12). So, for many new countries there has been a tendency to designate and locate a capital to either advance a national project, to create a balance between separate regional (and often national interests), or a combination of both.

Capitals are important symbols, and their location reflects how a new nation wants to see itself. With the emergence of new independent states out of existing colonies, the logic of capital location stood in contrast to existing colonial capitals. This had a lot to do with the different logics involved. For European colonizers, it was critical to establish the colony's primary city and capital in a major port. Therefore, colonial capitals such as Batavia, Lagos, Karachi, and Calcutta were developed to accommodate the imperial apparatus as well as provide the infrastructure for extraction. Any "national" interests of the colonial residents were immaterial. Besides which, many of the colonies included multiple populations brought together only because they fit within arbitrarily drawn lines.

Independence changed this logic dramatically as new states were compelled to designate a capital city. Very often, they would begin with the main city, which had served as the colonial center. But there were strong incentives to create entirely new capital cities. Table 8.1 shows the relocation of capital cities from a former to a new capital (taken from Schatz 2003 with updates by the authors). With the exception of Brazil, which became independent in 1822, these all represent newly independent post-colonial countries. This list is divided between the capitals established prior to the era of decolonization – following World War II – and those that came afterward.

Schatz (2003) provides a set of reasons why a country may wish to move a capital – often to create a whole new capital or to augment a smaller city with the infrastructure of a capital. There is simply the will of the leader – who often wants to exercise his (and it is almost always a "he") power by developing his own personally constructed city. There can also be rational reasons. A new capital might be more centrally located to the population, it will have more space and less congestion overall, and it will facilitate administrative efficiency. The third reason is the value of a capital city in the nationalist project: how well a new capital reflects the new nation and accomplishes certain goals of advancing the national project or mediating between rival cultural groups.

Capitals as Advancing a National Project

Many capitals are created as a way to optimistically reflect the nation and to point to where they see the future of the country. The creation of Brazil's national capital in Brasília represents the first instance of this phenomenon, at least in the post-colonial era.

Table 8.1 New and planned capital cities

Country	*Year*	*Former capital*	*New capital*
USA	1800	New York	Washington
Canada	1857	Toronto, Quebec City	Ottawa
New Zealand	1865	Auckland	Wellington
Japan	1868	Kyoto	Tokyo
Italy	1870	Turin, Florence	Rome
Turkey	1923	Istanbul	Ankara
Australia	1927	Melbourne	Canberra
Brazil	1956	Rio de Janeiro	Brasília
Pakistan	1959	Karachi	Islamabad
Botswana	1961	Mafeking	Gaberone
Libya	1963	Benghazi	Tripoli
Malawi	1965	Zomba	Liliongwe
Belize	1970	Belize City	Belmopan
Tanzania	1973	Dar es Salaam	Dodoma
Nigeria	1975	Lagos	Abuja
Côte d'Ivoire	1983	Abidjan	Yamoussoukro
Germany	1990	East Berlin, Bonn	Berlin
Kazakhstan	1997	Almaty	Astana
Malaysia	2000	Kuala Lumpur	Putrajaya
Myanmar	2005	Yangon (Rangoon)	Naypyidaw
Indonesia	2024	Jakarta	Nusantara*
South Korea	2030	Seoul	Sejong*

* planned

Brazil had long been an independent country but was very much part of the developing world – during the 1950s deemed the "third world" between Soviet and US influence. During the period as well, the bulk of Brazil's population was on the Atlantic coast, with a sparsely populated interior. This among one of the six largest countries in the world – the others being Russia, China, Canada, Australia, and the US.

The designation of Brasília was intended to demonstrate Brazil's national movement towards this interior. The acclaimed novelist John Dos Passos made several journeys into Brazil and authored a book called *Brazil on the Move*, published in 1963 (Brandt 2023). It was Dos Passos's view that Brazil represented an abundance of land, a belief in the future, and a movement to the interior. The new capital of Brasília was being established at the time of Dos Passos's travels, and Brazil's president at the time, Juscelino Kubitschek, declaims to him that:

> Brasília will constitute a goal, a place to head for on the high plateau. Building Brasília means roads. A movement of population into the fine farmlands of the interior is already going on … Brasília is the great goal of my administration.
>
> (Dos Passos, 1963, 69 in Brandt 2021)

The new capital was based on Brazil's future direction and orientation – and looked to the population to move to the new city in the central highlands with some jarringly modernistic architecture (Holford 1962) as epitomized in Figure 8.4.

About the same time, Pakistan leaders made the decision to create a new capital. The existing capital was Karachi. As was true of Rio de Janeiro in Brazil, Karachi was Pakistan's principal seaport on the Arabian Sea, its economic center, and one of its two largest cities. With the partition of India from Pakistan, it initially made sense to seat the government in Karachi. The other major city, Lahore, had been the center of Punjab and was seen as too proximate to India (Siddiqui 2012). But Karachi had been a colonial center, used by the British to funnel goods from the interior to shipment around the world. When Pakistan became independent after partition, Karachi also became the destination of most refugees from Hindu India – people who had little connection with Karachi region, but who exercised an outsized influence on its politics (Siddiqui 2012). What followed was a capital that represented national divisiveness rather than reflect the national character and aspirations.

Islamabad was proposed in 1959, 12 years after Pakistan was born and under the direction of a new leader, Ayub Khan. Disgusted with the ethnic and religious strife in Karachi, Khan proposed a new capital and landed on a brand-new space in the Himalayan foothills next to the existing city of Rawalpindi. This choice was made for a few reasons. Leaders were worried about how to coordinate the two "wings" of the new Pakistani state: east and west. So, they doubled down on the one basis of unity – the

Figure 8.4 Brasília's National Congress Building. This was the site of pro-Bolsonaro vandalism in January 2023
Source: Christoph Diewald

Muslim religion – and so calling the new capital the "City of Islam" reaffirmed these links (Siddiqui 2012). The location of the city was key as it pointedly ignored East Pakistan, with the rationale that a new Muslim capital should not be encircled by Hindus and Buddhists but instead would face west to the main centers of Islamic culture. Moreover, Islamabad's northern location was also considered more climatically salubrious, as Khan opined that hot and humid climates were not conducive to vitality and productivity (Harper 2010).

The principal idea behind the creation of Islamabad was an effort to create a stronger spatial focus for the nation. While the name Pakistan was an acrostic of several provincial names, Pakistan did not really possess a spatial identity. As Harper (2010, 79) describes it, Pakistan was a "pure and empty" idea, "emptied of history, culture, nature and even of religion, was the only idea that could encompass all Muslims and their different demands." As such, it did not possess the power of a movement mobilized around a distinct nation within a distinct territory. In fact, the elite of West Pakistan were primarily those refugees who fled from what was now majority-dominated India. They themselves had no roots in the new nation's soil. Islamabad would then become a clean, blank slate which would allow for an orderly center of administration and represent national unity instead of national discord (Harper 2010). As it happened, the new capital city reinforced the power of government and the dominance of the military in Pakistani life. And with the independence of East Pakistan in 1970, it had little chance to fulfill its stated goal of unifying the Pakistani people.

Capitals as Mediating between Cultural Groups

Especially with the formation of new multinational states, especially where the nations occupied a distinct area, the designation of capital cities was a geographically symbolic way to mediate between national regions. In the newly independent United States the location of Washington, DC was proposed as a compromise between the "north" and "south" regions. As Canada was on the verge of becoming independent, a pressing issue emerged as to where the capital city might be located. The Quebec province (Lower Canada) was mostly French whereas the Ontario province (Upper Canada) was mainly English speaking. The other provinces joining the original Confederation were also largely Anglophone but were too far removed from the country's pivot (Knight 1977).

The decision on where to locate a capital city truly went round and round before Canada's official independence in 1867 (Knight 1977). It was a choice made by members of the nascent assembly, but one needed to be sanctioned by the Crown. Quebec City and Kingston had been capitals of their respective provinces. Montreal and Toronto were the largest cities of these provinces and were also considered as good candidates at various times. There was even a period where alternating capitals between Quebec and Toronto seemed like the best approach. But Ottawa emerged as a natural compromise. It was centrally located within the population of the time (though perhaps a bit further to the north), it was far from US influence (of which the Canadians were rightfully anxious), and it was perceived as relatively neutral. Moreover, Ottawa was itself culturally mixed and was situated between the French and English dominant sections of Canada. While Ottawa did not command the fierce loyalty of any one faction, eventually it was acceptable to the majority.

More than a century later, similar considerations were used to decide on a capital for Nigeria. The colonial port of Lagos had served as a capital for 60 years, straddling the years when Nigeria was a British colony and after its independence in 1960. As a country, Nigeria exemplified many of the difficulties with colonies in Africa that gained independence after World War II. European powers had carved up these colonies to maximize resource extraction, with little consideration given to the cultural groups that lived within. Britain had actually merged two of its protectorates to create an amalgamated colony. This territory was large and contained over 200 ethnic groups. But it is also divided between three major groups: the Muslim Hausa in the north, the Christian Yoruba in the west, and the Christian Igbo in the east (Ibhawoh 2008). The British treated these groups differently and even in the run-up to independence there was clearly more pro-independence sentiment in the south than in the Muslim north (Honey 1999). The attempted secession of the Igbo east – as the new state of Biafra (see box in Chapter 7) – occurred just seven years after Nigerian independence.

Lagos was firmly in the Yoruba west, the most economically prosperous of the regions. Yet for many Nigerians, including those who were not Yoruban, this situation was fine. In fact, just prior to independence in the 1950s the Northern People's Congress (representing the Hausa) tried to retain Lagos as the capital but changed their position following independence (Moore 1984). Part of the reasoning was that congestion hindered the capital's growth – similar to the concerns in Karachi – and that the only way for Lagos to work as a capital was to "bulldoze the whole city and re-plan and re-build" (*Nigerian Tribune*, September 1971, quoted in Moore 1984, 172).

There was also a desire to see a new capital that would be in an unpopulated area, represent a neutral space, and be acceptable to the three major groups: Hausa, Yoruba, and Igbo (Abubakar 2014). Decision making was quick with a military government in charge. A commission was established to define a Federal Capital Region for Nigeria and the site was designated in the late 1970s. The capital region was located at the vertex of the three major national regions although Abuja – the new capital – came to be viewed as more of a northern city, since it was a few miles into Hausa territory (Moore 1984). Since the 1980s, the city has grown to seven million and has become a wealthy city in the Nigerian context (Abubakar 2014) (Figure 8.5).

The Urban Landscape and Nationalism

Chapters 3 and 4 point to how significant the landscape is to the national expression. This is especially apparent in the urban landscape. The greater density of the built environment in cities means that more humanly constructed structures will be found in the city. Also, since cities themselves symbolize the nation – serving as "crucibles of national identity and belonging" (Diener and Hagen 2018, 256) – they concentrate group identities. The urban landscape, particularly in a country's large and important cities, is displayed to visitors from the countryside, the urban residents, and to the rest of the world. It displays the sort of national identity the country's leaders wish to portray. These elites exert a great deal of influence over these symbols, but there are often counter narratives as well, offering alternative symbols of national identity within the same space.

National landscapes are often discursive, describing how members of a nation see themselves (Paasi 1999a). They are quite often banal and blended into the everyday – little realized by those passing by (Billig 1995). Cities include such elements of course, but

Figure 8.5 Abuja, Nigeria
Source: 3.29Fatima

they are also key repositories of "hot" nationalism, where national identity is flaunted. These elements of hot nationalism could be in the form of monuments, buildings, parks, transportation arteries, flags, and other decorations. And while such significant landscape elements also exist in smaller communities and in rural zones, the cities contain its grandest specimens.

Landscapes as Reflective of Shifting National Identities

The creation of the National Mall in Washington, DC exemplifies this assemblage of nationalist landscape elements. Washington was an intended capital city, built between the north and the south. The French architect of the city's design, Pierre L'Enfant, developed it in such a way as to project harmony between the street design, the incorporation of vistas, and the placement of important government buildings and other structures. In this, Washington fit within the "Grand Manner" of urban design which projected how the government wanted the nation to be portrayed. Throughout the 19th century, most of Washington was undeveloped – built in a swamp and with just a thin skeleton of government structures. During the 1850s and 1860s, many local monuments and statues were erected, but these tended to honor individual heroes and speak to more local identities (Savage 2009).

It was only after the Civil War that the National Mall, as a focal point of American national identity, began to take hold. The federal government expanded tremendously and so did the number of structures in official Washington (Savage 2009). The Washington Monument was finally built, and the McMillan Plan (adopted in 1902) established boundaries for the Mall as a separate national space. New monuments and buildings were constructed to adapt to this space and cherry trees were planted in and around the

Mall, becoming tourist destinations in themselves. Since then, grand buildings – many of them belonging to the Smithsonian Museum – have been constructed around the Mall and several new monuments – from the triumphal World War II monument to the subdued Vietnam Veterans Memorial – have been constructed within the Mall. In a sense, the National Mall depicts the complicated nationalism of the United States, while also presenting an officially sanctioned viewpoint. Yet one other aspect of the National Mall is also quite important: the use of the Mall as a center for demonstrations and protests. It allows people to contest issues and present rival conceptions of US identity.

The National Mall demonstrates both continuity and change in US national identity (Benton-Short 2016). The key structures that define the Mall – the national Capitol, the Washington Monument, and the Lincoln Memorial – remain in place. While aspects of their meaning have changed – we now may look at key figures more critically than in the past – the core of identity has remained fairly stable. At the same time, the change in the Mall reflects a shift in different sorts of buildings, monuments, and uses for the Mall.

No single monument reflected this better than the Vietnam Veterans Memorial (Figure 8.6), dedicated in 1982 and now the single most talked about, written about, and popular destination on the Mall (Savage 2009). It was anti-heroic, constructed as a low black wall, and was quite controversial from the outset. That should not have been surprising in that this memorial was built to address the painful history of a military defeat and for which there was no national consensus. It also went against expectations of what a war memorial should look like (Wagner-Pacifici and Schwartz 1991). Some critics called "a black gash of shame" (Garber 2007). Many memorials erected around the world affirm a pointed narrative. It may be one that affirms the triumph of the nation – supported by the whole of government and leadership. It may also represent an alternative view of history – a way of reshaping the narrative –as was done in the construction of so many Confederate monuments long after the Civil War had ended (O'Connell 2022). The Vietnam Veterans Memorial affirmed ambivalence, a striking departure from prior expressions of American identity.

For many countries, particularly those that moved away from colonialism and a more traditional economy, the urban landscape is an opportunity to present a whole new face to the state and to the nation. It goes to the heart of just what the nation is supposed to be, and the tension between rural and urban can be a big part of this.

The man most identified with India's independence, Mohandas Gandhi, viewed the country primarily as a network of villages, over half a million of them (Das and Choudhary 2018). He saw cities as necessary for some items but mostly as leeching off the rural peasantry, hearkening back to Thomas Jefferson who wanted the United States to be a society of yeoman farmers. Gandhi's vision developed when India was overwhelmingly rural and when the few large cities had been distorted by the needs of the British Empire and by the capitalist economy.

By contrast, India's first Prime Minister, Jawaharlal Nehru, cast the future of India in industrialization and urbanization. He was invested in building new cities and in renovating existing cities. One of the first examples of this was the city of Chandigarh, as the new capital of the Punjab. As with many capital cities, the development of Chandigarh could begin with a clean slate and provide a visible expression of India's potential as a newly empowered nation.

It was part of the 1950s ethos, but as with Islamabad, Chandigarh was designed by a Western architect. Nehru invited the famous French architect Le Corbusier to construct

Figure 8.6 Vietnam War Memorial in DC with the Washington Monument in the background
Source: Amy Davis

this city in the foothills of the Himalayan mountains. Le Corbusier designed a city with monumental buildings separated by open spaces (Figure 8.7). He established a grid system and divided the city into 29 self-contained sectors. Open spaces and a greenbelt were also established, reflecting Nehru's desire for a modern India (Das and Choudhary 2018). Other cities were later held up as representing the high technology nature of India. Bangalore (now Bengaluru), in the southern Deccan Plateau, emerged during the 1990s as a center of high technology growth (Kalra 2006). Westernized commercial retail, office buildings, malls, hotels, and especially housing for newly wealthy workers capped Bangalore's new landscape.

Landscapes as Reflective of Abrupt Changes

What happens when the core of "official" national identity changes abruptly? This has been the experience of many nations over the years but most recently has affected post-Communist regimes. Countries that were once part of the Soviet Union or in thrall to its power abruptly changed course in the period from 1989 to 1991. Each of these new or transformed countries made significant changes in their official monuments. Prague, Czech Republic took down a giant statue of Joseph Stalin; Vilnius, Lithuania removed all statuary honoring Soviet "heroes"; and countless cities toppled statues of Vladimir Lenin.

Figure 8.7 Chandigarh
Source: Fernando Stankuns

Ulan Bator, Mongolia demonstrates how this turnabout plays out in the public square (Myadar 2018). Mongolia is both an ancient empire, the birthplace of the Khans who swept through Asia and much of Europe, and a modern country that only came into being in 1921. Before this point, Mongolia had been a subject territory of the Chinese Empire and was especially grateful to the newly formed Soviet Union in securing its independence. The primary figure in Mongolia's early history was Damdiny Sukhbaatar, the original head of the Revolutionary Party and Mongolia's first leader. Sukhbaatar died young and became the personification of the Mongolian state and nation. Ulan Bator's central square was named after him and an enormous statue of him was erected in 1946. In 1954, his body was exhumed, and he was buried, along with Mongolia's second leader, in a giant mausoleum overlooking Sukhbaatar Square.

With the dissolution of the Soviet Union, Mongolia turned away from seven decades of Communism and began to reframe its identity (Myadar 2018). However, while the regime moved quickly to recast itself and its national identity, its monumental imprint moved at a slower pace. But in 2013, these changes came to a head. A statue of Lenin was brought down, a massive statue of Genghis Khan was put in place where the mausoleum had been, and the large central public plaza was renamed from Sukhbaatar to Genghis Khan Square. These reflected long standing official changes in Mongolia's national identity as Genghis Khan – the most powerful of the Medieval Khans – personified the new national identity.

The end of the Soviet Union forced changes throughout its considerable orbit. But like so many countries, Russia has grappled with the nature of its national identity. And this has been reflected in various landscape elements, particularly in its major cities. This was particularly true in the 1990s through early 2000s, as Russia was deciding which aspects of its past to remember and which to discard (Forest and Johnson 2002). All could agree on the glorious resistance to Nazism during the Great Patriotic War (World War II). The Russian Orthodox Church and Russian history before the Revolution were also fair game. Churches and monasteries were rebuilt, including a prominent cathedral that Stalin had destroyed. But post-Soviet Russia also disavowed the Communist Party and the Soviet economy. Most such museums and exhibits were neglected or repurposed. Most problematic were those monuments related to Lenin and the Russian Revolution. As in so many post-Communist cities, statues and monuments to Lenin epitomized the ambivalence. However, whereas Lenin could be considered a foreign interloper outside of Russia, his legacy was more nuanced within the country. Some Russians continued to adore the man and his memory. The Lenin Mausoleum, a massive structure on Red Square, continues to be open to this day under government support (Figure 8.8).

Landscapes as Reflective of Authoritarianism

The decision to advance a nationalist project can fuse with the desires of authoritarian leaders to project their version of nationalism. The excesses of the Nazi party in 1930s Germany led to widespread efforts, especially by Hitler's chief architect, Albert Speer, to remake the national landscape. The capital of Berlin was conceived as a series of massive plazas and streets, anchored by a seven kilometer long north-south boulevard that would pass monuments to war, soldiers, and Nazism (Hagen 2022a) (Figure 8.9). This reconstruction also required a tremendous amount of land, and Speer proposed obtaining much of it through evicting Jewish property owners, thus marrying the regime's interest in projecting a bombastic form of nationalism with its anti-Semitic goals (Jaskot 1996).

Figure 8.8 Lenin Mausoleum in Red Square, Moscow, Russia
Source: Brent Rostad

The Nazi regime also venerated the rural landscape as encapsulated in an iconic "beautiful village." These villages were seen as the essence of the German nation, devoid of alien elements, and cleaned up to exhibit the half-timbered architecture and an overall uniformity (Hagen 2022b).

During about the same timeframe, Italian fascists got in on the act as a second center – entitled the EUR District – was established five miles southeast of the city center. Initially it was meant to affirm Benito Mussolini's power with a series of monumental buildings. While construction was halted for a long time after Mussolini's overthrow, it was resumed in the 1960s and the area has a visual landscape that shows traces of its authoritarian beginnings (Agnew 1995).

The authoritarian landscape has been replicated across the world. One need only see such extreme examples as North Korea's Pyongyang. Here in the most privileged city in a country racked by poverty and fear, the landscape is constructed as a series of vast spaces, mostly empty except for periods of orchestrated demonstration. The landscape itself is marked by freakishly enormous buildings. The Ryugyong Hotel is 105 stories tall, made from concrete and in the shape of a pyramid because concrete construction requires that upper levels be lighter. Yet this pinnacle of the Pyongyang landscape has been unfinished since ground was broken in 1987 (Prisco 2020). The idea was that the Ryugyong Hotel would counter the architectural triumphs of South Korea. It now has the

Figure 8.9 Speer's Berlin plan
Source: Bundesarchiv, Bild 146III-373 / CC-BY-SA 3.0, CC BY-SA 3.0 DE, https://creativecommons.org/licenses/by-sa/3.0/de/deed.en, via Wikimedia Commons

distinction of being the tallest unoccupied building but shares with many authoritarian regimes the desire for pompous spectacle.

More recently, the dissolution of the Soviet Union led to several new countries. Among these, Kazakhstan strove to project a face of modernity and relative accord. This was symbolized in the movement of the capital away from Almaty to the new planned city of Astana, where it has been bound up with the longtime (now retired) president, Nursultan Nazarbayev (Koch 2010). Part of this projection of modernity has been the erection of gigantic facades, many of which are simply covers for existing Soviet-era buildings. To project tolerance and the idea of Kazakhstan as the crossroads of Eurasia, Nazarbayev commissioned an enormous pyramid that was designated the Palace of Peace and Accord (Koch 2010). Finally, the development of Astana has been entirely bound up with Nazarbayev's vision of himself and his personality cult. So much so that the capital was renamed "Nur-Sultan" upon his retirement in 2019, though the name was later changed back to "Astana" in 2022.

Urban Violence as Nationalistic Expression

Nationalism tends towards absolutism. Sometimes nationalist fervor erupts into violence against other groups, or the places inhabited by other groups. While several national groups may frame their differences within existing governance structures, "governance amidst severe and unresolved multicultural differences is viewed by at least one identifiable group in the city as artificial, imposed, or illegitimate" (Bollens 2007, 230). Within a city, this can take the form of extreme boundary marking and targeting of buildings symbolizing the presence of other identities. Acts of violence during periods of nationalist tensions are rarely "random"; they are usually intended.

The reason so much of this violence occurs in urban settings is because the city itself becomes a focal point of national tensions at higher scales. The boundaries between national groups are closely spaced and the city contains several structures of cultural and political significance (Bollens 2007). The urban density and built environment allow national terrorists to target places and peoples in crowded venues. Cities are also sites of ethnonationalist violence, in which one faction decides to wage war on those living within the same society but who are viewed by some as illegitimate. Sometimes the attacks are waged against the city itself and its representation as a cosmopolitan and multicultural place.

Sectarian Urban Violence

While rogue actors can and do play a role in urban violence, agents of the state and armies under sectarian control are more responsible. This was the sad case in Bosnia-Herzegovina, as it attempted to separate itself from what was left of Yugoslavia. Bosnia's main city, Sarajevo, became a poster child for the internecine violence that consumed Bosnia. But it was the violence in Mostar which demonstrated to the world the tragedy of nationalist conflict. Mostar was one of the major cities within Bosnia-Herzegovina. Built up during both the Ottoman and Austro-Hungarian Empires, it continued as a vital, multi-ethnic city within Tito's Yugoslavia. The Neretva River separated Mostar into an older city and a newer city. While there were some ethnic concentrations, there was also a fair bit of intermingling and conviviality as the diverse cultures accommodated one another and coexisted as members of civil society (Wise and Velayutham 2009).

With the advent of hostilities throughout Yugoslavia, Mostar became a flashpoint for conflict. The Yugoslavian army (at that point effectively the Serbian army) patrolled the streets and attacked the city. They forced all Muslim and Croatian citizens to abandon their homes in the older eastern part of the city and move to the western part. Yet the local Croat and Serb leaders made peace not long after, agreeing to divide Mostar at the expense of the Muslims. The Serbs withdrew but then Croat forces rounded up 15,000 Muslims, separating the able-bodied men from the women, children, and elderly who were forced to survive in the western part of the city.

The main commercial center of Mostar had been a broad boulevard paralleling the river. This became both a frontline and a virtual no-man's land between two sections of the city. The city's buildings were damaged, sometimes irretrievably, and the old Stari Most bridge was detonated. The result of this was a demographic sorting, as the once interspersed Croat and Muslim populations were almost completely partitioned (Bollens 2007; Calame and Charlesworth 2011).

Coward (2008) describes Mostar's destruction as an "urbicide." In this he argues that in addition to the targeting and dislocation of specific peoples, the perpetrators of violence also focused on the destruction of the city itself, at least those aspects of which they disapproved. As was true of Bosnia's capital of Sarajevo, nationalist Serbs and Croats abhorred what the large cities represented. The built environment of these cities symbolized their diversity. So, the leveling of buildings and streets to rubble was more than simply collateral damage but part of the destruction in the symbolic landscape, in those elements that represented the heterogeneous Yugoslavia of old. The mix of mosques, churches, libraries, and museums spoke to the ideal of a plural society, and this was anathema to the ethnonationalist forces.

Ethnonationalist Violence Against the City

Ethnonationalism bases national belonging on cultural homogeneity, with ethnonationalists defining the criteria. People who do not fit into their conception are considered suspect and not really part of the nation. This can happen in countries of all make-ups but is especially acute in countries with a sizable proportion of ethnic minorities. Some of the more virulent ethnonationalist movements have targeted these minority groups, even if members consider themselves also to belong to the nation.

Cities are often seen as centers of the sort of diversity and tolerance that ethnonationalist movements abhor. Intellectuals marshal the resentment of rural and small-town residents against the urban cosmopolitan elite. Shaw (2004) cites several examples of this phenomenon, ranging from the wars waged against urban residents by the Communist regimes of Stalin's Soviet Union, Mao's China, and the Khmer Rouge's Cambodia – where the city was kept intact but many residents eliminated, to the Croat and Serb nationalist leaders' contempt of diverse Yugoslav cities, such as Sarajevo and Mostar, where the cities themselves were destroyed. While many of these leaders also waged war against their rural peasantry, it was often the cities that came up for special scrutiny. Removing the population or destroying their ability to live within the city was often the goal. In this case "urbicide" destroys not just cities but the heterogeneity represented by cities. Shaw goes as far as to equate thus "urbicide" with actual "genocide" since the destruction of place also leads to the eradication of a group.

We can find an example of this phenomenon in the United States itself, with the 1995 Oklahoma City bombing. When the Alfred Murrah Federal Building was destroyed, killing 168 people, the initial suspicion was cast towards Muslim terrorists. However, it turned out that the attack was perpetrated by Timothy McVeigh, an anti-Semitic white nationalist who was inspired by a book that advocated the overthrow of the government and the eradication of all minority groups. McVeigh was radicalized by the sieges at Ruby Ridge and Waco in the early 1990s and finally decided to attack a federal building. He was soon caught and later executed. But McVeigh continued to believe that his actions were justified.

Oklahoma City represented firstly an attack on the government, but it was very much connected to the white power movement. The federal building within Oklahoma City represented what would now be described as the "globalist" state, in service to a large diverse community. McVeigh's motivations were clearly influenced by the white power movement to which he subscribed (Belew 2018).

Attacks on federal installations can be one means of expressing both anti-government hatred and hatred at the nation as it is. Ethnonationalists wish to return to an earlier version of national identity which excludes diverse groups. The state is seen as an enabler of such diversity and so is often targeted. Other symbols of diversity have also been attacked. More recently, mass shootings took place in a Pittsburgh synagogue and in a Black church in Charlestown. Both were perpetrated by white supremacists.

Conclusion

While nationalism tends to be considered at the scale of the country, nationalism manifests itself often in cities and towns, often exposing power relations at the local scale. Cities are national seats of power, signaling to those inside and outside of the nation the direction of national interest and power. Elites live in cities, these cities tend to be seats of government and economic power, and they generate well known, or at least stereotypical identities of their own. While cities are flashpoints of national activity and power, rural places are usually thought of as holding the "true" content of the nation. Nationalists and ethnonationalists often define themselves in opposition to the city, claiming a different national identity as tied to a rural lifestyle and social contract. The tension between a city and rural hinterlands often acts as both a hot and banal thermometer for the health of the nation overall.

Cities are cosmopolitan places of cultural, ethnic, and linguistic diversity. As such, they can highlight a plethora of national identities in the very place meant to be a singular national power. Some cities have attempted to repress national differences by constructing walls, segregating groups, or even expelling minorities. Sometimes these groups evolve in tandem, forming pockets of different national identities in one city which coexist and add to the diverse urban fabric. Other times, tensions flare, and conflict arises in city streets as groups fight for more recognition, mobility, and power in the city that they call home. Even the location of capital cities can be used in accommodating national groups, or to project the nation into a future frontier.

From national malls to village squares, nationalism is on display in the local landscape. Capital cities display national moments with monumental structures to triumph or pensive memorials to tragedy or controversy. Towns also contain national memorials, albeit on a smaller scale. While cities are constantly holding court for protests, celebrations, and demonstrations, towns contain spaces for these public displays as well. These performances are often held in the same spaces where national identity is inscribed upon the landscape. These landscapes and the ways that people interact with them demonstrate national past and present, showing both continuity and change in place.

Chapter 9

How Do Nationalism and Geography Interact?

For a series entitled the *Key Ideas in Geography*, a discussion of nationalism and national identity would seem to be a perfect fit. While it is true that many of the seminal concepts related to nationalism have come from other fields – notably political science, sociology, history, and philosophy – any discussion of the nationalism phenomenon rests on certain key geographical principles. This is because, quite simply, nationalism is intrinsically geographical, which means that it is impossible to conceive of nationalism without a geographical basis. Other forms of political affiliation have relied more closely on personal relationships, often structured, and codified over the ages, but still tied to relationships of power.

Yet nationalism relies on bonding to a specific place. This place is stretched to encompass an area, often delimited, which is then taken to be both a rightful possession of the members of the nation and a symbol of nationhood itself. Nationalism is also applied to the bonding of a people to each other, and not as simply a vertical relationship to a personal sovereign. When nationalism stepped in as a political ideology, the notion of lands as the private demesne of lords and kings was swept away to one where these lands belonged to an entire people. Despite the bewildering range of nationalist expressions over the centuries, with huge variations in size, population, composition, and control – not to mention disagreements over the nation's true extent – this fundamental feature remains constant.

Territory

The significance of territory to human society far predates nationalism. It predates humanity itself as it is a feature of much animal behavior. Humans used territory to restrict space. The earliest cities from Sumer civilization were walled, offering both a defensive benefit but also clearly demarcating city dwellers from the peasantry outside the city. Pre-nationalist political organization also relied heavily on territorial control. There were many short-lived empires, but the genius of the Roman Empire in the West and the Han, Sui, and Tang empires in China lay in their ability to carve out and administer territory – provinces in Rome, circuits in China. In these instances, territorial control was still a top-down process, with power emanating from the emperor to governors to administrators. All successful empires and larger kingdoms copied this model.

Many would argue that true nationalism was predicated on the origin of a state system. It created a change in mentality. As Murphy (1996) argues, the mid-17th-century Peace of Westphalia ushered in a state system of mutually agreed upon territories and accepted state boundaries. This gradually changed the view from one where territorial

DOI: 10.4324/9780429435782-9

control overlapped in a variety of complex ways to a more straightforward system where political, social, and economic life could be based around discrete territorial units. This sovereign state system was still a long time in the making, but it eventually prevailed, not only in Europe but throughout the rest of the world, as a product of European influence.

Nationalism later harnessed this flat territorial organization to the idea that national peoples, rather than kings and queens, were the rightful owners of this territory. Sovereignty was no longer vested in a person but in a national people. In turn, the rightful national territory was to be determined not as a plot owned by some princeling or duke, but as it was occupied – possessed – by a particular people. That is why Fichte's addresses to the German nation were so powerful. He was exhorting the entirety of the German people and the lands they occupied, this when Germany was broken into dozens of small sovereign parcels.

Territorial nationalism became an important principle in the development of 19th-century nation-states, especially Germany and Italy, but also in the Americas, where new states were being formed under nationalist ideals. Arguments in South America to split away from Spain rested on arguments of their distinctive culture (Lynch 2001). Still many multinational empires remained. The first decades of the 20th century affirmed the primacy of the nation-state ideal. World War I rejiggered the European map to accommodate a set of brand-new nation-states. World War II prompted the end of overseas colonialism and the development of about 100 new independent states, under the premise that these would be based around the promise of a distinct nationhood.

This model of territorial nation-states begets two distinct problems. The first problem has to do with the disjuncture between state and nation. Every political geographer realizes that the nation-state ideal is far from the reality. Yet this ideal sends a powerful message. It affects the way we speak of states and nations – culminating in entities such as the United Nations, which is really an association of sovereign states (see Figure 9.1). It also puts the onus on states to "build" a nation within the state's territorial confines, even if no single nation predominated before. The European model was exported to the world, but Europe itself includes numerous instances where nations do not fit neatly into states (Williams 1982). Globally, the tension in reconciling several thousand nations into some 200 states leads to tremendous unrest as many national aspirations remain unfulfilled.

The second problem involves the fixity of the territorial state itself. While the world considers these individual areas to be inviolable, states are far from unitary, sealed actors on the world stage. In what Agnew (1994) terms the "territorial trap," states are viewed as the primary spatial containers of social, economic, and political life. In fact, flows of capital, people, ideas, movements, culture, and all manner of human life course through and around state boundaries. State boundaries are still important but are also quite porous. Transnationalism is a clear manifestation of the multifarious flows which occur between countries, and between localities within these countries. People enter from specific places, sometimes legally, sometimes not, and they will continue to maintain ties with the places they came from. Often the movements back and forth occur several times a year, as people come in order to gain economic benefit, visit family, or improve their prospects. Such flows can also have a strong political dimension, as displaced populations may form a diaspora – continuing to maintain strong relations with their homelands and intending to return someday. Nationalism is geographical, but not always based on fixed spaces; it can also be based on flows.

Figure 9.1 UN Geneva Headquarters
Source: Cayla

Absolute sovereignty within the state territory is challenged as well, as regions rise and command their own sense of loyalty and affinity. Regional identity is a significant counter to the ideal of the singular nation-state. Paasi (2003) distinguishes between those aspects that constitute the distinctive character of a region and may be used by residents to distinguish one region from another, and the sort of regional consciousness which can rival loyalty with the larger state. These regional identities may nest within this broader loyalty, but that is not always the case (Herb and Kaplan 1999). This too shows the power of the nationalist ideal, as regions are not simply administrative apparatuses meant to govern large states but are places with their own sense of affinity to other regional inhabitants and to the land of the region itself.

The territoriality of a nation-state relies as well on its boundaries, a topic long studied in political geography (see Minghi 1963). If the state is a "container" then the boundary provides the political casing. The evidence on the landscape is clear, with guarded entry points, customs checks, and sometimes even walls! Borders are significant but are tremendously misleading. While they represent the official geopolitical order, they are usually quite messy in real life. (Sometimes when international boundaries produce numerous enclaves and exclaves, they are administratively messy as well.) Borders cut across cultural communities, they are themselves regions which display the intermingling of two or more nations. In many instances, borders themselves form a region of hybridity

(Kaplan and Häkli 2002). In several cases, large national regions straddle several borders, as is the case of Kurdistan (Dahlman and Moradi 2018).

Landscape

Geographical interest in landscape has a long provenance. After all, the landscape is the field of play, the visible (and sometimes invisible) manifestation of physical and human activity. What we observe on the landscape reflects underlying processes, but landscapes can be agents in their own right. The turn towards nationalism changed the composition of the human landscape. Of course, memorials to God and King continued, but were now accompanied by symbols of the nation itself.

Landscapes represent both the national territory and the totality of the nation. The admittedly blurry distinction between a hot nationalism and a more everyday, banal nationalism can be witnessed on the ground. Countries adorn their lands with all manners of memorials: to the epic events that have transpired and to the people who have served the nation well. Nowhere is this more evident than in the capital city of a country where statuary, monuments, and hallowed spaces proliferate. Ulan Bator's square honoring Genghis Khan, Paris's Place de la République, and Washington's National Mall are all public spaces meant to glorify the nation. These are also spaces which contain many symbols of national identity. The showcase in the Place de la République is not an historical person but is instead the French nation personified in the figure of Marianne. For the United States, the Statue of Liberty (given by France) serves much the same purpose. It humanizes one desired attribute of the American nation.

Expressions of hot nationalism are also subject to alteration, sometimes erasure. These changes can be made by outsiders, seeking to harm the nation, or by members of the nation itself. The term "urbicide" describes how parts of a city might be obliterated, with attention to specific symbols of the national project (Coward 2008).

From the viewpoint of national insiders, certain symbols of nationalism may become inconvenient and then be removed. After the fall of Communism in Eastern Europe and then the former Soviet Union itself, statues of Vladimir Lenin were taken down in city after city. For countries which had been under Soviet sway, Lenin represented an interesting "national" figure. He symbolized the suzerainty of Communist ideology over nationalist aspirations. So, Lenin's removal from the public landscape reaffirmed each country's nationalism. Genghis Khan Square in Ulan Bator reasserted the past Mongolian identity as all traces of Lenin were removed. Within Russia itself, Lenin's legacy was bound up with a particular Russian nationalism. Here, Lenin and the huge Lenin Mausoleum have remained. In the United States, there has been a longstanding push to remove symbols of the Confederacy. Ironically, these were mostly installed well after the South lost the Civil War to assert the continued value of this variety of Southern nationalism and to instill an oppressive racial order. With a greater awareness of the racist legacies embodied in such Confederate symbols, the push has been to remove these where possible, though this has met with fierce resistance from those who see this as an obliteration of Southern heritage.

A national landscape is populated with countless denotations of national identity. Edensor (2020) points out those commonplace symbols which subtly convey national culture. The types of housing styles, the churches, and other institutions, how stores are arranged, the adaptation of nature, the linguistic landscape: all of these are critical markers which effectively flag the landscape (Lowenthal 1994). Sometimes these everyday images become iconic in their own right, such as the Dutch windmill and polder. Other times, they may strike the observer as oddly out of place – such as Tyrolean villages in northern Italy – until the national context is better understood.

It is within this everyday landscape that alternative nationalisms might play out. Nationalists may see a landscape that poorly represents them. If such groups gain power, they can try to amend the landscape by law. Quebec's 1977 language law did just that, requiring that all signage in the province (except for that of the Federal government) only be in French. If they do not yet have the power to amend the landscape, they might take matters into their own hands. Signage or structures which so clearly convey undesired dominance of a nation, such as road signs, may be removed or defaced.

The landscape itself adumbrates the national project. Yet this landscape is amplified by other tokens of the nation. Stamps and coins are produced by the government and are in ubiquitous usage. Such items purvey what the government wants to display. Very often this depicts a landscape of some sort, of either nature or the built environment. With regime change comes a change in what is placed on a stamp. The movement from the Soviet Union to Russia sparked a change from internationalist to more inward Russian ethnic themes, with a great deal of emphasis on nature and heritage (Brunn 2011).

Scale

The terms "nationalism" and "national identity" presume a particular geographical scale – bigger than a district, much more expansive than a locality, yet normally smaller than a continent or a broad world region. The term "nation" also presumes a particular societal scale – bigger than a clan, much more than a family, yet smaller than a broad "civilization." The extent of nations clearly differs. Russia occupies a vast expanse from Europe to Siberia. Malta is just as much a nation as Russia but covers only 122 square miles, one eighth the size of Rhode Island. If considered in terms of population, such "nations" as China can be thousands of times larger than the innumerable small nations found throughout the earth. Yet while absolute scale in such instances clearly varies, relational scale remains the same – ensconced in the middle, between the local and the global.

Circumstances led to a world where people envisioned themselves as belonging to this middle ground of a nation. The primacy of the nation, and its occupance of a middle scale, is something recent. Certainly, other political identities did not and do not exist at this scale. A 15th-century observer of the modern world would not recognize our arrangement. The territoriality of states had yet to be established as tribes, feudal manors, city-states, chiefdoms, and empires prevailed. These all occupied many scales from the familial to the civilizational.

This scalar vision of nations masks the degree to which the national scale is challenged from below and from above. In every country, the dominant nation conceals significant regional variation. It is far from uniform but varies according to geography and society. Some regionalism is very much nationalism in disguise, actuating strong national separatist movements (Williams 1982). But a lot of regionalism exhibits much lower levels of intensity, and regional feeling does not necessarily follow administrative geographies (Antonsich 2010). Some states are successful in accommodating these variations, whereas others are not. The other issue lies in the degree that residents connect well beyond the national polity. The most convenient reason is globalization, which has diminished the salience of national scale activities while heightening international networks (Brenner 1999).

The fixed scale of a nation-state assumes that all loyalties, sovereignties, and legitimacies accrue within that single territorial "container." But we know that this does not apply to the actual world. States are porous, nations bleed into one another, and state and nation are rarely aligned. There is another aspect regarding how well people belonging to one nation can spread their loyalty to other political and cultural entities. The clearest instances of this may be found within the European Union. It is not as if "Europe" becomes just another nation in its own right, but it incorporates many aspects of the political authority of a state with the cultural cohesion of a nation. In the late 1990s, the further development of Europe's political apparatus created a framework for various actions and one that prompted increasing European identification, at least among citizens of some countries (Murphy 1999). Other supranational entities are less developed, but we can find numerous instances where loyalty and feeling go well beyond the national scale.

At times, the supranational element exists in the absence of any clear nation. At what scales do diasporas function? Some are visible offshoots of a territorial homeland, but others may engender national identity from members of the diaspora themselves. Chinese populations dispersed within Southeast Asia have often been met with ambivalence or even outright hostility. Their own sense of distinctiveness has likewise complicated an easy coexistence. In this realm, the supranational identity is oriented around the Chinese diaspora population, based both on continued economic networks but also a sense of identification with the Nanyang Chinese rather than the Chinese living within Mainland China (Pan 1994).

This highlights how we might reconsider scale itself. The notion of a nation as a fixed territorially based entity presupposes what Häkli (2018) describes as a "flat ontology" where the geopolitical world is conceived as a set of horizontal spaces and where the national scale neatly exists along a continuum from the tiniest locality to the entirety of the earth (and beyond). The nature of human activity does not always follow such clean pathways. Instead, connections are drawn in all directions and across many dimensions. The activities of a particular national group may be reflected at multiple scales, all at the same time. Affinities and affiliations do not neatly map onto our partitioned world.

Mapping

The flat ontology that orders the geopolitical world and defines our national consciousness stems from and is bolstered by the ubiquity and power of maps. Maps illustrate the extent of a nation, yet they do so in simplified terms. As two-dimensional objects that portray a three-dimensional world, maps are necessarily distorted, simplified, and generalized (Monmonier 2018). Yet maps, with their problematic inconsistencies, help us understand the world, and offer comfort among the high degree of cultural and political confusion faced in the real world. As banal ephemera which appear in a multitude of contexts and medias, maps are taken for granted as a reflection of reality, when geographers know this to be far from the truth. However, this is done on purpose, as the simplified nature of maps reflects the simplified ideal of a homogeneous nation-state.

Maps are very much taken as immortal facts. Although the world and its political borders have changed immensely over time, new versions of maps sweep the past away, re-drawing and re-coloring reality for new generations of viewers. The most current versions of maps obscure, even obliterate the past by portraying a cartographer's current point of view and communicating an immobile set of geographical "truths" via paper or screen. Especially today, when tools like Google Maps are so heavily depended upon and so frequently referenced, maps hold an incredible amount of power in terms of how we order space and see ourselves in the world. The world political map shows blocks of color bounded by solid lines of black. Within each of these blocks of color lies a ubiquitous nation, homogeneous in identity, territoriality, and history, as uniform as the color which portrays it. On the one hand maps show a story of who is inside and who is outside of the nation. The borders both within and surrounding a state show a history of the violent process of nation building, and much can be gleaned from the presences and absences in maps. Yet on the other hand, these colors and lines ignore the messy result of such projects. Border communities, minority groups, stateless nations, indigenous lands are all obscured by the maps we reference every day. These maps are not for decision making. Rather, they are used to portray a political project and reflect a nation's place.

Nationalism can be thought of as the sense of a nation's place in the world. This "place" is both literal in terms of site and situation, but also figurative, in terms of power, status, and wealth. Maps reflect both these literal and figurative senses of nationalism. The power of a country might be portrayed in terms of relative size, which is easily distorted through a projection choice. Military power might be symbolized using powerful primary colors, or even illustration symbols. These understandings of nations and nationalism are both established and reinforced through maps. Nationhood is a political project, and maps are reflective of the dominant socio-spatial narratives that stem from this project (Harley 2002). As state governments change the conceptions of national meaning, so too do maps change. In Latin America, for example, as states have incorporated multi-ethnic recognition into the national narrative and governance model, the production, distribution, and use of maps recognizing indigenous and Afro-Caribbean peoples have proliferated (Radcliffe 2010), albeit unevenly. Maps, as a core state product or as a product of protest, offer a means to understand and explore norms and values, and ideals of nations.

While state-sanctioned maps that appear in media, textbooks, and in the classroom are central to the solidification and normalization of a "core" or state-sanctioned national identity, maps are also powerful tools of protest and critical reimagining among marginalized groups. Sub-state nations that do not identify with the core of the nation-state ideal rarely receive cartographic recognition within the solid block of color portraying the political and social world. So, these groups create their own maps in protest of the "reality" that mainstream maps portray.

Nationalism and Geography into the Future

Though experts differ as to its origins, we can confidently say that nationalism has been a coherent and powerful ideology for at least 200 years. There is little within the present moment that suggests the potency of nations will wither in the foreseeable future. This is partly due to nationalism's flexibility. It coexists comfortably with all manner of economic ideology, from unfettered free-trade capitalism to conditions approaching autarchy – North Korea is among the most nationalistic countries on the planet! Nationalism fits within a range of political systems, from democratic states to those ruled by dictators. Nationalism may be more obvious in authoritarian regimes, but it also permeates the fabric of societies which are largely democratic.

Nationalism is a constant presence, yet there are several forces which could profoundly change its contours. With every passing year, the world becomes more interconnected. Increases in global media access, travel, and permanent migration supported human connection that transcends borders, allowing for realization of mutual experiences (Cox 2021). The global economy is one such force, creating a need among many countries to merge to develop larger markets. But the profound impact of a global media system means that currents that come to prominence in one country echo across many countries. While many for a while thought that such global interconnections could result in a more cosmopolitan outlook, all different sorts of movements can spread. Our current partitioned news environment means that many people are exposed only to highly polarized information. We would be foolish to think that only messages of international cooperation and tolerance for diversity will spread.

But while globalization has affected nationalism, it is unlikely to diminish it. Economic exchange reduces the importance of borders in commerce but increases nationalist identification due to changes in domestic economic systems (Murphy 2013). The reaction to globalization can be profoundly felt, as some countries have reversed their affiliation in these larger supranational entities, notably Britain's exit from the European Union, culminating in a wave of globalization backlash. Globalization might also, ironically, feed its supposed antithesis. Leaders like Jair Bolsonaro in Brazil, Donald Trump in the US, and Giorgia Meloni in Italy are representatives of a new iteration of far-right populism and anti-cosmopolitanism. These conservative populist leaders have harnessed a nationalistic sentiment which recycles anti-global rhetoric from the interwar period. On the heels of Trump's failed coup, many United States conservatives have looked to national leaders such as Viktor Orbán of Hungary who embodies what they view as desirable political traits, aspiring to a political system which rejects democratic choice in favor of a strong man (Beauchamp 2022). The expansion of right-wing populism, ethnonationalist outlooks, anti-immigrant sentiment – these have long existed under the surface but seemed to be having a moment, creating powerful currents in the United States, India, and Russia to name some of the more prominent examples. Will these sorts of movements change the complexion of nationalism, leading it to become xenophobic and less inclusive? Or will they be met with a strong counter-reaction as we have seen in some other countries?

Changes in the physical environment could represent the greatest existential threat to humanity in the 21st century. The consequences of anthropomorphically induced climate change have been confirmed with growing intensity since it first emerged as an issue in the 1980s. Every scientist expects there to be significant environmental alterations because of atmospheric carbon, but the precise contours are still unknown. What we do know is that the impacts will fall more heavily on the poorest countries, both because poor societies often have precarious infrastructure and high poverty levels, and because many of these countries are in places where climate change may be at its most intense. Areas that suffer widespread drought, or unbearable levels of heat, or food shortages: these are precisely those areas likely to no longer be able to accommodate existing populations. The result will be a wave of climate refugees, desperate to find places where they can live (Berchin et al. 2017). Most international refugee policy focuses only on political persecution while the overwhelming number of new migrants will arrive because they simply cannot live in their home. How countries deal with these refugees will stretch their understanding of nationalism. The current streams of migrants, and the nationalist reactions that they have provoked, may seem like only a trickle when compared to what lies ahead.

Changes in technology will also impact our sense of nations. Given the continued importance of maps to nationalism and national identity, new forms of mapping signal a shift in the popular conception of nations and nation-states. New tools, platforms, and mapping designs have proliferated within a primarily online-based community of mappers. Tools like OpenStreetMap, Google Earth, and ArcGIS online, coupled with social media platforms like Reddit, have generated amateur products that challenge and can change socio-spatial norms of what a nation is. Maps are no longer products designed and generated exclusively by the state. Rather, they can reflect multiple points of view and disseminate through virtual spaces where people spend significant amounts of time engaging in work and recreation.

What will come to be known as the Great Pandemic of 2020–22 has upended lives in too many ways to enumerate. The question is how many of these changes are temporary and how many will be long-lasting? One change forced by pandemic-required social distancing has been the explosion of virtuality in meetings, work, associations, and other modes of communication. When COVID hit, the computer infrastructure in many countries was robust and advanced enough to allow for easy virtuality. Fifteen years earlier and such options would not have been available. Easy digital communication obliterates distance between space and allows people to retain ties more easily across the globe. When Vertovec (2004) discussed the importance of cheap, international calls, or Brinkerhoff (2009) propounded the value of digital diasporas, they were referring to a larger phenomenon that will supercharge within the next several decades. Virtual communication is not as good as the real thing – we have sadly discovered this – but it comes closer than ever before. With complete virtual reality, it will be even more possible for people to live rich virtual lives.

What are the implications of virtual space for nationalism? Transnationalism relies on the retention of these ties. It connects people. Nationalism is clearly rooted in geography but the virtual or even imagined geography of the internet may offer a compelling substitute – even more so with continued technological improvements. As economics and politics liberalized the movement of people, things, and ideas, diasporas and transnational populations have been able to tap into a network of kin and generate virtual spaces of national camaraderie like never before. Rather than make territory obsolete, these virtual spaces remind people of connections, products, and people that come from specific places (Rosecrance, 1996). The importance of these digital diasporas may prove ever more important as populations of more developed countries decline. States can embrace these diaspora and transnational populations (Délano and Gamlen 2014). The utilization of virtual spaces and tools will be a lynchpin for nationalist goals.

Nationalism is a philosophy, an ideology, and an imagined family. It is something that subliminally pervades people's lives but which – under certain conditions – becomes an emotional maelstrom, motivating people to perform acts they would not otherwise do. Nationalism is also quite flexible. As scholars have long pointed out, there are as many kinds of nationalisms as there are nations – setting off a variety that reaches well into the thousands. And even within the same nation, the meaning of nationalism takes many guises, each of which is in continuous flux. Yet it is within its very ubiquity that nationalism shows its profound influence over human affairs. All of us, caught up in this particular historical moment, would be hard-pressed to imagine a world where nationalism fades away. Such would be a world so different from the present that all the anchors guiding our political lives would also vanish.

Bibliography

Abbink, J. 2011. Ethnic-based federalism and ethnicity in Ethiopia: reassessing the experiment after 20 years. *Journal of Eastern African Studies* 5 (4), pp. 596–618.

Abubakar, I. 2014. Abuja city profile. *Cities* 41, pp. 81–91.

Aburish, S. 2004. *Nasser: The Last Arab.* New York, NY: St. Martin's Press.

Adeney, K. 2017. Does ethnofederalism explain the success of Indian federalism? *India Review* 16 (1), pp. 125–148.

Agnew, J. 1994. The territorial trap: the geographical assumptions of international relations theory. *Review of International Political Economy* 1 (1), pp. 53–80.

Agnew, J. 1995. *Rome.* New York, NY: John Wiley and Sons.

Agnew, J. 2002. The "civilisational" roots of European national boundaries. In D. Kaplan and J. Hakli (eds.), *Boundaries and Place: European Borderlands in Geographical Context.* Lanham, MD: Rowman & Littlefield.

Agnew, J. 2017. *Globalization and Sovereignty: Beyond the Territorial Trap.* Lanham, MD: Rowman & Littlefield.

Agrawal, A., and Kumar, V. 2017. Cartographic conflicts within a union: finding land for Nagaland in India. *Political Geography* 61, pp. 123–147.

Aitken, S., and Craine, J. 2011. Affective geovisualisations. In M. Dodge, R. Kitchin, and C. Perkins (eds.), *The Map Reader: Theories of Mapping Practice and Cartographic Representation*, pp. 278–280. Hoboken, NJ: John Wiley & Sons.

Ajami, F. 1978. The end of pan-Arabism. *Foreign Affairs* 57, pp. 355–373.

Alcock, A. 1992. The protection of regional cultural minorities and the process of European integration: the example of South Tyrol. *International Relations* 11 (1), pp. 17–36.

Alderman, D. H. 2010. Surrogation and the politics of remembering slavery in Savannah, Georgia (USA). *Journal of Historical Geography* 36 (1), pp. 90–101.

Alderman, D. H. 2016. Place, naming and the interpretation of cultural landscapes. In B. Graham and P. Howard (eds.), *The Ashgate Research Companion to Heritage and Identity*, pp. 195–213. London: Routledge.

Alderman, D. H. 2017. Review of Tales from the Haunted South: Dark Tourism and Memories of Slavery by Tiya Miles. *African American Review* 50 (1), pp. 87–89.

Alderman, D. H., and Modlin, E. A. 2014. The historical geography of racialized landscapes. In C. Colten and G. Buckley (eds.), *North American Odyssey: Historical Geographies for the Twenty-first Century*, 273–290. Lanham, MD: Rowman & Littlefield.

Aljazeera. 2022. Donetsk and Luhansk: What to know about Ukraine's rebel regions. Aljazeera.

Allport, G. W. 1979. *The Nature of Prejudice.* New York, NY: Perseus Books Publishing.

Álvarez, B. 2017. Transitional justice in settled democracies: Northern Ireland and the Basque Country in comparative perspective. *Critical Studies on Terrorism* 10 (3), pp. 542–561.

Americans and Their Flag. 2016. *New York Times*, September 1.

Anderson, B. 1991. *Imagined Communities: Reflections on the Origins and Spread of Nationalism.* London: Verso.

Anderson, B. 1999. Indonesian nationalism today and in the future. *Indonesia* 67, pp. 1–11.

Andersson, H., and Smith, A. 2021. Flags and fields: a comparative analysis of national identity in butter packaging in Sweden and the UK. *Social Semiotics*, pp. 1–22.

Anthias, P. 2019. Ambivalent cartographies: exploring the legacies of indigenous land titling through participatory mapping. *Critique of Anthropology* 39 (2), pp. 222–242.

Antonsich, M. 2010. Exploring the correspondence between regional forms of governance and regional identity: the case of Western Europe. *European Urban and Regional Studies* 17 (3), pp. 261–276.

Applegate, C. 2008. Germany. In G. Herb and D. Kaplan (eds), *Nations and Nationalism: A Global Historical Overview.* Santa Barbara: ABC-CLIO.

Archdeacon, T. J. 1984. *Becoming American.* New York, NY: Simon and Schuster.

Archetti, E. 2007. Masculinity, primitivism, and power: gaúcho, tango, and the shaping of Argentine national identity. *Gender, Sexuality, and Power in Latin America Since Independence* 3, 212.

Archetti, E. P. 2017. Masculinity and football: the formation of national identity in Argentina. In *Game Without Frontiers*, pp. 225–243. London: Routledge.

Ardrey, R. 1967. *The Territorial Imperative.* London: Primrose Hill Books.

Argelaguet, J. 2021. The relevance of language as a predictor of the will for independence in Catalonia in 1996 and 2020. *Politics and Governance* 9 (4), pp. 426–438.

Armstrong, J. 1982. *Nations Before Nationalism.* Chapel Hill, NC: University of North Carolina Press.

Azaryahu, M., and Golan, A. 2004. Zionist homelandscapes (and their constitution) in Israeli geography. *Social & Cultural Geography* 5 (3), pp. 497–513.

Bairner, A. 2008. Sport, nationalism and globalization: relevance, impact, consequences. *Hitotsubashi Journal of Arts and Sciences* 49 (1), pp. 43–53.

Bale, J. 2002. *Sports Geography.* London: Routledge.

Bale, J., and Sang, J. 2013. *Kenyan Running: Movement Culture, Geography and Global Change.* London: Routledge.

Bar-Gal, Y. 1991. *The Good and the Bad: A Hundred Years of Zionist Images in Geography Textbooks.* Queen Mary and Westfield College, University of London, Department of Geography.

Bar-Gal, Y. 2003. *Propaganda and Zionist Education: The Jewish National Fund, 1924–1947.* Rochester, NY: University Rochester Press.

Barkey, E., and Gavrilis, G. 2015. The Ottoman millet system: non-territorial autonomy and its contemporary legacy. *Ethnopolitics* 15 (1), doi:10.1080/17449057.2015.1101845.

Bartkus, V. 1999. *The Dynamic of Secession.* Cambridge, UK: Cambridge University Press.

Bassett, T. J. 1994. Cartography and empire building in nineteenth-century West Africa. *Geographical Review* 84 (3), pp. 316–335.

Bassett, T. J. 2017. Signs of the times: commercial road mapping and national identity in South Africa. In J. R. Akerman (ed.), *Decolonizing the Map: Cartography from Colony to Nation.* Chicago, IL: University of Chicago Press. pp. 339–376.

Batuman, B. 2010. The shape of the nation: visual production of nationalism through maps in Turkey. *Political Geography* 29 (4), pp. 220–234.

Beaubian, J., Lonsdorf, K. and Maynes, C. 2022. Occupied regions of Ukraine vote to join Russia in staged referendums. National Public Radio, September 27. www.npr.org/2022/09/27/1125322026/russia-ukraine-referendums.

Beauchamp, Z. 2022. Ron DeSantis is following a trail blazed by a Hungarian authoritarian. *Vox*, April 28.

Bekus, N. 2010. Nationalism and socialism: the Soviet case. In *Struggle over Identity: The Official and the Alternative "Belarusianness".* Central European University Press. Retrieved from http://books.openedition.org/ceup/602.

Belew, K. 2018. *Bring the War Home: The White Power Movement and Paramilitary America*. Cambridge, MA: Harvard University Press.

Bell, C. 1997. The "real" New Zealand: rural mythologies perpetuated and commodified. *The Social Science Journal* 34 (2), pp. 145–158.

Beller, S. 2015. *Antisemitism: A Very Short Introduction*. New York, NY: Oxford University Press.

Bendix, R. 1992. National sentiment in the enactment and discourse of Swiss political ritual. *American Ethnologist* 19, pp. 769–790.

Benner, E. 2013. Nationalism: intellectual origins. In J. Breuilly (ed.), *The Oxford Handbook of the History of Nationalism*, pp. 36–55. Oxford, UK: Oxford University Press.

Bennett, A., and Woodward, I. 2016. Festival spaces, identity, experience and belonging. In J. Taylor (ed.), *The Festivalization of Culture*, pp. 25–40. London: Routledge.

Benton-Short, L. 2016. *The National Mall: No Ordinary Public Space*. Toronto: University of Toronto Press.

Benvenisti, M. 2000. *Sacred Landscape: The Buried History of the Holy Land Since 1948*. Berkeley, CA: University of California Press.

Benwell, M. 2014. Considering nationality and performativity: undertaking research across the geopolitical divide in the Falkland Islands and Argentina. *Area* 46 (2), pp. 163–169.

Benwell, M. 2016. Encountering geopolitical pasts in the present: young people's everyday engagements with memory in the Falkland Islands. *Transactions of the Institute of British Geographers* 41 (2), pp. 121–133.

Benwell, M. C., and Pinkerton, A. 2020. Everyday invasions: Fuckland, geopolitics, and the (re) production of insecurity in the Falkland Islands. *Environment and Planning C: Politics and Space* 38 (6), pp. 998–1016.

Benwell, M. *et al.* 2021. Stitching together the nation's fabric during the Chile uprisings: towards an alter-geopolitics of flags and everyday nationalism. *Geoforum* 122, pp. 22–31.

Berchin, I., Valduga, I., Garcia, J. and Guerra, J. 2017. Climate change and forced migrations: an effort towards recognizing climate refugees. *Geoforum* 84, pp. 147–150.

Berg, C. 2008. Politics, not sport, is the purpose of the Olympic Games. *Institute of Public Affairs Review: A Quarterly Review of Politics and Public Affairs* 60 (3), pp. 14–18.

Berlin, I. 1976. *Vico and Herder: Two Studies in the History of Ideas*. New York: Vintage Books.

Berríos Martínez, R. 1977. Independence for Puerto Rico: the only solution. *Foreign Affairs* 55, p. 561.

Beswick, J. E. 2007. *Regional Nationalism in Spain*. Bristol, UK: Multilingual Matters.

Bhabha, H. K. 1994. *The Location of Culture*. London: Routledge.

Bhargava, R. 2012. Should Indian federalism be called multinational? In M. Seymour and A. Gagnon (eds.), *Multinational Federalism*. New York, NY: Palgrave Macmillan.

Billig, M. 1995. *Banal Nationalism*. London: Sage.

Black, J. 2000. *Maps and Politics*. Chicago: University of Chicago Press.

Blair, P. 2003. *An Introduction to Anglo-Saxon England*. Cambridge, UK: Cambridge University Press.

Blaney, D. L. and Inayatullah, N. 2000. The Westphalian deferral. *International Studies Review* 2 (2), pp. 29–64.

Boal, F. 1969. Territoriality on the Shankill-Falls divide, Belfast. *Irish Geography* 6 (1), pp. 30–50.

Bockelman, B. 2011. Between the gaucho and the tango: popular songs and the shifting landscape of modern Argentine identity, 1895–1915. *The American Historical Review* 116 (3), pp. 577–601.

Bollens, S. 2007. Urban governance at the nationalist divide: coping with group-based claims. *Journal of Urban Affairs* 29 (3), pp. 229–253.

Bonds, A., and Inwood, J. 2016. Beyond white privilege: geographies of white supremacy and settler colonialism. *Progress in Human Geography* 40 (6), pp. 715–733.

Boomgaard, M. 2008. The rise of militant Bretonite. *National Identities* 10 (3), pp. 281–293.

Bowden, B. 2003. Nationalism and cosmopolitanism: irreconcilable differences or possible bedfellows? *National Identities* 5 (3), pp. 235–249.

Bowden, B. 2008. Postcolonial nationalist philosophies. In G. Herb and D. Kaplan (eds.), *Nations and Nationalism: A Global Historical Overview*. Santa Barbara: ABC-CLIO.

Bradfield, S. 2006. Separatism or status-quo?: Indigenous affairs from the birth of land rights to the death of ATSIC. *Australian Journal of Politics & History* 52 (1), pp. 80–97.

Branch, J. 2013. *The Cartographic State: Maps, Territory, and the Origins of Sovereignty*. Cambridge, UK: Cambridge University Press.

Brandt, S. T. 2023. The Brazilian scene: David Lowenthal, John Dos Passos, and the importance of "scene" and Brazil to geographic inquiry. *Geographical Review*, 113 (2), pp. 171–190.

Brenner, N. 1999. Beyond state-centrism? Space, territoriality, and geographical scale in globalization studies. *Theory and Society* 28 (1), pp. 39–78.

Breton, R. 1964. Institutional completeness of ethnic communities and the personal relations of immigrants. *American Journal of Sociology* 70 (2), pp. 193–205.

Breton, R. 1988. From ethnic to civic nationalism: English Canada and Quebec. *Ethnic and Racial Studies* 11 (1), pp. 85–102.

Breuilly, J. 1985. Reflections on nationalism. *Philosophy of the Social Sciences* 15 (1), pp. 65–75.

Breuilly, J. 1993. *Nationalism and the State*. Manchester, UK: Manchester University Press.

Breuilly, J. 2013. Introduction: concepts, approaches, theories. In J. Breuilly (ed.), *The Oxford Handbook of the History of Nationalism*. Oxford, UK: Oxford University Press.

Brickell, K., and Datta, A. (eds.). 2011. *Translocal Geographies*. Farnham, UK: Ashgate Publishing, Ltd.

Brinkerhoff, J. 2009. *Digital Diasporas: Identity and Transnational Engagement*. Cambridge, UK: Cambridge University Press.

British Broadcasting Corporation News. 2014. Scottish referendum: Scotland votes "No" to independence. September 14. www.bbc.com/news/uk-scotland-29270441.

British-Irish Agreement Act. 1999. Electronic Irish Statute Book. www.irishstatutebook.ie/eli/1999/act/1/enacted/en/html.

Brohm, J. M. 1978. *Sport, a Prison of Measured Time: Essays*. Henlow, UK: Ink Links Limited.

Bronner, S. 2002. *Folk Nation: Folklore in the Creation of American Tradition*. Lanham, MD: Rowman & Littlefield Publishers.

Broschek, J. 2016. Federalism in Europe, America and Africa: a comparative analysis. In W. Hofmeister and E. Tayoa (eds.), *Federalism and Decentralization: Perceptions for Political and Institutional Reforms*, pp. 23–50. Singapore: Konrad-Adenauer-Stiftung.

Brosius, C., and Polit, K. M. (eds.). 2020. *Ritual, Heritage and Identity: The Politics of Culture and Performance in a Globalised World*. Milton Park, UK: Taylor & Francis.

Brown, D. 1998. Why is the nation-state so vulnerable to ethnic nationalism? *Nations and Nationalism* 4 (1), pp. 1–15.

Brunn, S. D. 2000. Stamps as iconography: celebrating the independence of new European and Central Asian states. *GeoJournal* 52(4), pp. 315–323.

Brunn, S. D. 2011. Stamps as messengers of political transition. *Geographical Review* 101 (1), pp. 19–36.

Bryan, J. 2011. Walking the line: participatory mapping, indigenous rights, and neoliberalism. *Geoforum* 42 (1), pp. 40–50.

Bryson, L. 1987. Sport and the maintenance of masculine hegemony. In *Women's Studies International Forum* 10 (4), pp. 349–360. Oxford, UK: Pergamon.

Bull, H. 1959. What is the Commonwealth? *World Politics* 11 (4), pp. 577–587.

Burke, P. 2013. *Social History of Knowledge: From Gutenberg to Diderot*. Hoboken, NY: John Wiley & Sons.

Butler, J. 1997. *Excitable Speech: A Politics of the Performative*. New York, NY: Routledge.

Calame, J., and Charlesworth, E. 2011. *Divided Cities*. Philadelphia, PA: University of Pennsylvania Press.

Calder, K. 2021. *Global Political Cities: Actors and Arenas of Influence in International Affairs*. Washington, DC: Brookings Institution Press.

Campos-Delgado, A. 2018. Counter-mapping migration: irregular migrants' stories through cognitive mapping. *Mobilities* 13 (4), pp. 488–504.

Caney, S. 1998. National self-determination and national secession. In P. Lehning (ed.), *Theories of Secession*, pp. 149–178. London: Routledge.

Carment, D. 2003. Assessing state failure: implications for theory and policy. *Third World Quarterly* 24 (3), pp. 407–427.

Carrión, J. 2006. The war of the flags: conflicting national loyalties in a modern colonial situation. *CENTRO Journal* 18 (2), pp. 100–123.

Caryl, C. 2015. The world's next country. *Foreign Policy*, January 21, 2015.

Casserly, R. 2018. Localizing the national: performing British identity in Northern Ireland. In S. Reily and K. Brucher (eds.), *The Routledge Companion to the Study of Local Musicking*, pp. 429–441. London: Routledge.

Castonguay, C. 1987. The Anglicization of Canada, 1971–1981. *Language Problems and Language Planning* 11, pp. 22–34.

Chacko, E., 2015. Hybrid sensibilities: highly skilled Asian Indians negotiating identity in private and public spaces of Washington, DC. *Journal of Cultural Geography* 32 (1), pp. 115–128.

Chatterjee, P. 1993. *Nationalist Thought and the Colonial World: A Derivative Discourse*. London: Zed Books.

Chazan, N. (ed.). 1991. *Irredentism and International Politics*. Boulder, CO: Lynne Rienner Publishers.

Cheesman, N. 2017. How in Myanmar "national races" came to surpass citizenship and exclude Rohingya. *Journal of Contemporary Asia* 47 (3), pp. 461–483.

Chun, A. 1996. Discourses of identity in the changing spaces of public culture in Taiwan, Hong Kong and Singapore. *Theory, Culture & Society* 13 (1), pp. 51–75.

Church, J. 2011. Review of Lydia L. Moland's *Hegel on Political Identity: Patriotism, Nationality, Cosmopolitanism*. *Notre Dame Philosophical Reviews* https://ndpr.nd.edu/reviews/hegel-on-political-identity-patriotism-nationality-cosmopolitanism.

Clark, J. C. 2000. Protestantism, nationalism, and national identity, 1660–1832. *The Historical Journal* 43 (1), pp. 249–276.

Coakley, J. 2016. Introduction: dispersed minorities and non-territorial autonomy. *Ethnopolitics* 15 (1), pp. 1–23.

Cohen, A. 2020. *The Politics of Elite Culture*. Berkeley, CA: University of California Press.

Cohen, B. 2018. *The Geography of Money*. Ithaca, NY: Cornell University Press.

Cohen, R. 1996. Diasporas and the nation-state: from victims to challengers. *International Affairs* 72 (3), pp. 507–520.

Cohen, R. 2008. *Global Diasporas: An Introduction*. London: Routledge.

Collier, S. 1983. Nationality, nationalism and supranationalism in the writings of Simon Bolivar. *Hispanic American Historical Review* 63 (1), pp. 37–64.

Conforti, Y. 2010. East and west in Jewish nationalism: conflicting types in the Zionist vision? *Nations and Nationalism* 16 (2), pp. 201–219.

Connor, W. 1972. Nation-building or nation-destroying? *World Politics* 24 (3), pp. 319–355.

Connor, W. 1978. A nation is a nation, is a state, is an ethnic group is a… *Ethnic and Racial Studies* 1 (4), pp. 377–400.

Connor, W. 1993. Beyond reason: the nature of the ethnonational bond. *Ethnic and Racial Studies* 16 (3), pp. 373–389.

Connor, W. 2018. *Ethnonationalism*. Princeton, NJ: Princeton University Press.

Conversi, D. 1990. Language or race?: The choice of core values in the development of Catalan and Basque nationalisms. *Ethnic and Racial Studies* 13 (1), pp. 50–70.

Conversi, D. 2000. *The Basques, the Catalans, and Spain: Alternative Routes to Nationalist Mobilisation*. Reno, NV: University of Nevada Press.

Conversi, D. 2004. *Ethnonationalism in the Contemporary World: Walker Connor and the Study of Nationalism*. Milton Park, UK: Psychology Press.

Conversi, D. 2006. Genocide, ethnic cleansing and nationalism. In G. Delanty and K. Kumar (eds.), *The SAGE Handbook of Nations and Nationalism*, pp. 320–334. London, UK: Sage.

Conzen, M. 1993. Culture regions, homelands, and ethnic archipelagos in the United States: methodological considerations. *Journal of Cultural Geography* 13 (2), pp. 13–29.

Cook, M. 2016. Counter-narratives of slavery in the Deep South: the politics of empathy along and beyond River Road. *Journal of Heritage Tourism* 11 (3), pp. 290–308.

Cornell, S. 2015. Processes of native nationhood: the indigenous politics of self-government. *The International Indigenous Policy Journal* 6 (4).

Cosentino, D. 2015. Designing the identity of a new state: Kosovo flag. *The Design Journal* 18 (4), pp. 607–612.

Côté, I. 2012. Autonomy and ethnic diversity: the case of Xinjiang Uighur Autonomous Region in China. In A. Gagnon and M. Keating (eds.), *Political Autonomy and Divided Societies: Imagining Democratic Alternatives in Complex Settings*, pp. 171–184. London: Palgrave Macmillan.

Coward, M. 2008. *Urbicide: The Politics of Urban Destruction*. London: Routledge.

Cox, L. 2021. *Nationalism: Themes, Theories, and Controversies*. Singapore: Springer Singapore.

Crampton, J. 2006. The cartographic calculation of space: race mapping and the Balkans at the Paris Peace Conference of 1919. *Social & Cultural Geography* 7 (5), pp. 731–752.

Crampton, J. 2009. Cartography: performative, participatory, political. *Progress in Human Geography* 33 (6), pp. 840–848.

Cronon, W. 2011. *Changes in the Land: Indians, Colonists, and the Ecology of New England*. New York, NY: Hill and Wang.

Crystal, D. 1987. *The Cambridge Encyclopedia of Language*. Cambridge: Cambridge University Press.

Culcasi, K. 2010. Locating Kurdistan: contextualizing the region's ambiguous boundaries. In A. Diener and J. Hagen (eds.), *Borderlines and Borderlands: Political Oddities at the Edge of the Nation-state*. Lanham, MD: Rowman & Littlefield Publishers.

Culcasi, K. 2011. Cartographies of supranationalism: creating and silencing territories in the "Arab Homeland". *Political Geography* 30 (8), pp. 417–428.

Culcasi, K. 2012. Mapping the Middle East from within: (counter-) cartographies of an imperialist construction. *Antipode* 44 (4), pp. 1099–1118.

Culcasi, K. 2016. Warm nationalism: mapping and imagining the Jordanian nation. *Political Geography* 54, pp. 7–20.

Culcasi, K. 2018. Imagining the Arab Homeland and Palestine. In G. Herb and D. Kaplan, (eds.), *Scaling Identities: Nationalism and Territoriality*, pp. 137–156. Lanham, MD: Rowman & Littlefield.

Cullingford, E. B. 1990. "Thinking of Her... as... Ireland": Yeats, Pearse and Heaney. *Textual Practice* 4 (1), pp. 1–21.

Dahbour, O., and Ishay, M. (eds.). 1995. *The Nationalism Reader*. Lanham, MD: Humanity Books.

Dahlman, C. and S. Moradi. 2018. Partition and national fragmentation of Kurdistan. In G. Herb and D. Kaplan (eds), *Scaling Identities: Nationalism and Territoriality*, pp. 223–238. Lanham, MD: Rowman & Littlefield.

Dalin, C. 2004. The Great Wall of China. In P. Ganster and D. Lorey (eds.), *Borders and Border Politics in a Globalizing World*. Lanham, MD: Rowman & Littlefield Publishers.

Dalrymple, W. 2015. The great divide. *The New Yorker* 29.

Das, D., and Choudhary, B. 2018. From "Rural" to "Urban" India. In A. Diener and J. Hagen (eds.), *The City as Power: Urban Space, Place, and National Identity*, pp. 221–234. Lanham, MD: Rowman & Littlefield Publishers.

Das, T., Bhattacharyya, R. and Sarma, P. 2020. Revisiting geographies of nationalism and national identity in Bangladesh. *GeoJournal* 87, pp. 1–22.

Davis, E. 2008. Uyghur Muslim ethnic separatism in Xinjiang, China. *Asian Affairs: An American Review* 35 (1), pp. 15–30.

Dawn, E. 1991. The origins of Arab nationalism. In R. Khalidi, M. Muslih, and R. Simon (eds.), *The Origins of Arab Nationalism*, pp. 3–30. New York, NY: Columbia University Press.

De Zamaróczy, N. 2017. Are we what we play? Global politics in historical strategy computer games. *International Studies Perspectives* 18 (2), pp. 155–174.

Dear, M., and Burridge, A. 2005. Cultural integration and hybridization at the United States-Mexico borderlands. *Cahiers de géographie du Québec* 49 (138), pp. 301–318.

Del Barrio, M. J., Devesa, M., and Herrero, L. C. 2012. Evaluating intangible cultural heritage: the case of cultural festivals. *City, Culture and Society* 3 (4), pp. 235–244.

Del Casino, V. J., and Hanna, S. P. 2005. Beyond the "binaries": a methodological intervention for interrogating maps as representational practices. *ACME: An International Journal for Critical Geographies* 4 (1), pp. 34–56.

Délano, A., and Gamlen, A. 2014. Comparing and theorizing state–diaspora relations. *Political Geography* 41, pp. 43–53.

Della Porta, D., and Portos, M. 2021. A bourgeois story? The class basis of Catalan independentism. *Territory, Politics, Governance* 9 (3), pp. 391–411.

Deutsch, K. 1966. *Nationalism and Social Communication*. Cambridge, MA: The MIT Press.

Devasundaram, A. 2018. Beyond brand Bollywood: alternative articulations of geopolitical discourse in new Indian films. In R. Saunders and V. Strukov (eds.), *Popular Geopolitics*, pp. 152–173. London: Routledge.

Diener, A., and Hagen, J. 2018. *The City as Power: Urban Space, Place, and National Identity*. Lanham MD: Rowman & Littlefield.

Dittmer, J. 2009. Captain America and Captain Britain. In R. Weiner (ed.), *Geopolitical Identity and the Special Relationship: Captain America and the Struggle of the Superhero. Critical Essays*, pp. 135–146. Jefferson, North Carolina and London: McFarland & Company.

Dittmer, J. 2012. *Captain America and the Nationalist Superhero: Metaphors, Narratives, and Geopolitics*. Philadelphia, PA: Temple University Press.

Dittmer, J., and Bos, D. 2019. *Popular Culture, Geopolitics, and Identity*. Lanham, MD: Rowman & Littlefield.

Dixon, C. 2008. Haiti. In G. Herb and D. Kaplan (eds.), *Nations and Nationalism: A Global Historical Overview*. Santa Barbara: ABC-CLIO.

Dodds K. 1993. Geopolitics, cartography and the state in South America. *Political Geography* 12 (4), pp. 361–381.

Dodds, K. 2006. Popular geopolitics and audience dispositions: James Bond and the internet movie database (IMDb). *Transactions of the Institute of British Geographers* 31(2), pp. 116–130.

Dodds, K. 2008. Hollywood and the popular geopolitics of the War on Terror. *Third World Quarterly* 29 (8), pp. 1621–1637.

Dodds, K. 2010. Jason Bourne: gender, geopolitics, and contemporary representations of national security. *Journal of Popular Film & Television* 38 (1), pp. 21–33.

Dodds, K. 2011. Gender, geopolitics, and geosurveillance in *The Bourne Ultimatum*. *Geographical Review* 101 (1), pp. 88–105.

Dodds, K., and Hemmings, A. D. 2014. Recent developments in relations between the United Kingdom and the Argentine Republic in the South Atlantic/Antarctic region. *Polar Record* 50 (2), pp. 119–127.

Dodge, M., Kitchin, R., and Perkins, C. 2011. *Rethinking Maps: New Frontiers in Cartographic Theory*. London: Routledge.

Doob, L. 1976. *Patriotism and Nationalism: Their Psychological Foundations*. Westport, CT: Greenwood Press.

D'Souza, C. 1994. The breakup of Yugoslavia. *Economic and Political Weekly* 29 (48), pp. 3027–3031.

Dupree, N. 2002. Cultural heritage and national identity in Afghanistan. *Third World Quarterly* 23 (5), pp. 977–989.

Durante, C. and Loftus, J. 2016. The Bechdel Cast. iheartRadio. www.iheart.com/podcast/105-the-bechdel-cast-30089535.

Duruji, M. 2012. Resurgent ethno-nationalism and the renewed demand for Biafra in south-east Nigeria. *National Identities* 14 (4), pp. 329.

Eagleton, T. 1991. *Ideology: An Introduction*. London: Verso.

Eckert, A. 2013. Anti-Western doctrines of nationalism. In J. Breuilly (ed.), *The Oxford Handbook of The History of Nationalism*, pp. 56–74. Oxford, UK: Oxford University Press.

Edensor, T. 2004. Automobility and national identity: representation, geography and driving practice. *Theory, Culture & Society 21* (4–5), pp. 101–120.

Edensor, T. 2020. *National Identity, Popular Culture and Everyday Life*. London: Routledge.

Edney, M. 2009. *Mapping an Empire: The Geographical Construction of British India, 1765–1843*. Chicago, IL: University of Chicago Press.

Elden, S. 2010. Land, terrain, territory. *Progress in Human Geography* 34 (6), pp. 799–817.

Eley, G. and Suny, R. 1996. Introduction: from the moment of social history to the work of cultural representation. In G. Eley and R. Suny (eds.), *Becoming National: A Reader*, pp. 3–37. New York: Oxford University Press.

Elgenius, G. 2018. *Symbols of Nations and Nationalism: Celebrating Nationhood*. Dordrecht: Springer.

Enze, H. 2011. From domestic to international: the politics of ethnic identity in Xinjiang and Inner Mongolia. *Nationalities Papers: The Journal of Nationalism and Ethnicity* 39 (6), pp. 941–962.

Ergin, M. 2014. The racialization of Kurdish identity in Turkey. *Ethnic and Racial Studies* 37 (2), pp. 322–341.

Esman, M. 1986. The Chinese diaspora in Southeast Asia. In G. Sheffer (ed.), *Modern Diasporas in International Politics*. New York: St. Martin's Press.

European Central Bank. 2021. The Euro. Retrieved 19 August 2021, from www.ecb.europa.eu/euro/html/index.en.html.

European Union. 2021. Values and Identities of EU citizens. *Report Eurobarometer* 508.

Evered, K. 2008. Turkey. In G. Herb and D. Kaplan (eds.), *Nations and Nationalism: A Global Historical Overview*. Santa Barbara: ABC-CLIO.

Fagan, M. and Gubbala, S. 2022. Positive views of European Union reach new highs in many countries. Pew Research Center, October 13.

Falser, M. 2014. From a colonial reinvention to postcolonial heritage and a global commodity: performing and re-enacting Angkor Wat and the Royal Khmer Ballet. *International Journal of Heritage Studies* 20 (7–8), pp. 702–723.

Fessha, Y. 2017. The original sin of Ethiopian federalism. *Ethnopolitics* 16 (3), pp. 232–245.

Fichte, J. G. 1807/2008. *Fichte: Addresses to the German Nation*. Cambridge, UK: Cambridge University Press.

Fincher, R., Iveson, K., Leitner, H. and Preston, V. 2014. Planning in the multicultural city: celebrating diversity or reinforcing difference? *Progress in Planning* 92, pp. 1–55.

Forest, B., and Johnson, J. 2002. Unraveling the threads of history: Soviet-era monuments and post-Soviet national identity in Moscow. *Annals of the Association of American Geographers* 92 (3), pp. 524–547.

Forslid, R. and Nyberg, S. 2020. Brexit and the survival of the Union. In A. Bakardjieva Engelbrekt, K. Leijon, A. Michalski and L. Oxelheim (eds.), *The European Union and the Return of the Nation State: Interdisciplinary European Studies*. New York, NY: Palgrave Macmillan.

Forsyth, M. 2007. Federalism, nationality, statehood: the problem of the European Union. In M. Burgess and J. Pinder (eds.), *Multinational Federations*, pp. 162–191. London: Routledge.

Fortna, B. 2005. Change in the school maps of the late Ottoman Empire. *Imago Mundi* 57 (1), pp. 23–34.

Fountaine, S. 2020. Themes of connection and progress in rural television: New Zealand's Country Calendar 1990–2015. *Media International Australia* 174 (1), pp. 109–124.

Fountaine, S., and Bulmer, S. 2022. Telling stories about farming: mediated authenticity and New Zealand's country calendar. *Television & New Media* 23 (1), pp. 81–99.

Fournier, P. 2021. La souveraineté tient bon. *L'actualité*, February 25. https://lactualite.com/politique/la-souverainete-tient-bon.

Fraunhar, A. 2005. Mulata Cubana: the problematics of national allegory. In L. Shaw and S. Dennison (eds.), *Latin American Cinema: Essays on Modernity, Gender, and National Identity*, 160–179. Jefferson, NC: McFarland.

Freeman, M. 1998. The priority of function over structure. In P. Lehning (ed.), *Theories of Secession* 13. London: Routledge.

Frew, E., and White, L. 2011. *Tourism and National Identity: An International Perspective*. Milton Park, UK: Taylor & Francis.

Fruchter-Ronen, I. 2013. The Palestinian issue as constructed in Jordanian school textbooks, 1964–94: changes in the national narrative. *Middle Eastern Studies* 49 (2), pp. 280–295.

Fung, A. 2001. What makes the local? A brief consideration of the rejuvenation of Hong Kong identity. *Cultural Studies* 15 (3–4), 591–601.

Garber, K. 2007. A milestone for a memorial that has touched millions. *US News and World Report*, November 3. www.usnews.com/news/articles/2007/11/03/milestone-for-a-memorial-that-has-touched.

Gautier, L. 2018. *Puerto Rico: Independence, Industrial Policy, and Growth*. Lanham, MD: Lexington Books.

Gawrys, M. 2021. *The Implications of Changing Border Structure: A Case Study in Kosovo*. Master's thesis, Kent State University.

Gellner, E. 2003. *Nations and Nationalism*. Ithaca, NY: Cornell University Press.

Genberg, D. 2003. Borders and boundaries in post-war Beirut. In F. Colombjon and A. Erdentug (eds.), *Urban Ethnic Encounters: The Spatial Consequences*, pp. 97–112. London: Routledge.

Gershoni, I. 1997. Rethinking the formation of Arab nationalism in the Middle East, 1920–1945: old and new narratives. In J. Jankowski and I. Gershoni (eds.), *Rethinking Nationalism in the Arab Middle East*, pp. 3–25. New York: Colombia University.

Getz, D. 2010. The nature and scope of festival studies. *International Journal of Event Management Research* 5 (1), pp. 1–47.

Gilbert, E. 1999. Forging a national currency. In E. Gilbert and E. Helleiner (eds.), *Nation-states and Money: The Past, Present and Future of National Currencies*, pp. 25–46. New York, NY: Routledge.

Gilbert, P. 2008. Philosophy, national character and nationalism, 1880–1945. In G. Herb and D. Kaplan (eds.), *Nations and Nationalism: A Global Historical Overview*. Santa Barbara: ABC-CLIO.

Gitelman, Z. 1998. The decline of the diaspora Jewish nation: boundaries, content, and Jewish identity. *Jewish Social Studies* 4 (2), pp. 112–132.

Glassheim, E. 2008. Ethnic cleansing and genocide. In G. Herb and D. Kaplan (eds.), *Nations and Nationalism: A Global Historical Overview*. Santa Barbara, CA: ABC-CLIO

Golan, A. 2008. Israel. In G. Herb and D. Kaplan (eds.), *Nations and Nationalism: A Global Historical Overview*. Santa Barbara, CA: ABC-CLIO.

Goldfarb, M. 2009. *Emancipation: How Liberating Europe's Jews from the Ghetto Led to Revolution and Renaissance*. New York, NY: Simon and Schuster.

Gomez, C. 2021. Social solidarity as a dimension of transitional justice: the case of cartography and identification of mass graves in post-conflict Colombia. *The International Journal of Human Rights* 26 (1), pp. 1–24.

Gottmann, J. 1973. *The Significance of Territory*. Charlottesville: University Press of Virginia.

Gottmann, J. 1975. The evolution of the concept of territory. *Social Science Information* 14 (3), pp. 29–47.

Gounden, V. 2010. Reflections on identity conflicts and nation building in contemporary South Africa. *South Asian Journal of Peacebuilding* 3 (2), pp. 1–5.

Gow, J. and Carmichael, C. 2000. *Slovenia and the Slovenes: A Small State and the New Europe*. Bloomington, IN: Indiana University Press.

Greenfeld, L. 1992. *Nationalism: Five Roads to Modernity*. Cambridge: Harvard University Press.

Greiner, C., and P. Sakdapolrak 2013. Translocality: concepts, applications and emerging research perspectives. *Geography Compass* 7 (5), pp. 373–384.

Griffiths, R. 2016. *The Age of Secession*. Cambridge, UK: Cambridge University Press.

Griffiths, R. D. and Butcher, C. R. 2013. Introducing the international system (s) dataset (ISD), 1816–2011. *International Interactions* 39 (5), pp. 748–768.

Guibernau, M. 2013. *Nationalisms: The Nation-state and Nationalism in the Twentieth Century*. Hoboken, NJ: John Wiley & Sons.

Guibernau, M. 2014. Prospects for an independent Catalonia. *International Journal of Politics, Culture, and Society* 27 (1), pp. 5–23.

Gunderman, H. 2017. Blurring the protagonist/antagonist binary through a geopolitics of peace: Star Trek's Cardassians, antagonists of the Alpha Quadrant. *The Geographical Bulletin* 58(1), pp. 51.

Habtu, A. 2004. Ethnic pluralism as an organizing principle of the Ethiopian Federation. *Dialectical Anthropology* 28 (2), pp. 91–123.

Hagen, J. 2022a. Places of memory, historic preservation, and place attachment in Nazi Germany. *Geographical Review* 112 (1), pp. 147–170.

Hagen, J. 2022b. Geographies of totalitarianism in village, town and countryside: architecture, landscape design and spatial planning in Nazi Germany. In N. Koch (eds.), *Spatializing Authoritarianism*, pp. 61–84. Syracuse, NY: Syracuse University Press.

Häkli, J. 1999. Cultures of demarcation: territory and national identity in Finland. In G. Herb and D. Kaplan (eds.), *Nested Identities: Nationalism, Territory, and Scale*, pp. 123–150. Lanham, MD: Rowman & Littlefield.

Häkli, J. 2018. Afterword: transcending scale? In G. Herb and D. Kaplan (eds.), *Scaling Identities: Nationalism and Territoriality*, pp. 271–282. Lanham, MD: Rowman & Littlefield.

Hale, E. 1863. The man without a country. *The Atlantic Monthly*, December 1863.

Hall, P. 1998. *Cities in Civilization*. New York, NY: Pantheon Books.

Halmai, G. 2018. Fidesz and faith: ethno-nationalism in Hungary. *Verfassungsblog: On Matters Constitutional*. https://verfassungsblog.de/fidesz-and-faith-ethno-nationalism-in-hungary/.

Han, E. 2011. From domestic to international: the politics of ethnic identity in Xinjiang and Inner Mongolia. *Nationalities Papers* 39 (6), pp. 941–962.

Hancock, L. 2016. Into the IRIS: a model for analyzing identity dynamics in conflict. *National Identities* 18 (4), pp. 409–423.

Hannum, K. 2016. *Sociolinguistic Geographies in Galicia, Spain*. MA Thesis, Kent State University.

Hannum, K. 2020. Minority language groups in the state system. In S. Brunn and R. Kehrein (eds.), *Handbook of the Changing World Language Map*. Cham: Springer.

Hannum, K. 2020. Language policy and practice between urban and rural schools in Galicia, Spain. *GeoJournal* 85 (4), pp. 1009–1024.

Hannum, K. and Rhodes, M. 2018. Public art as public pedagogy: memorial landscapes of the Cambodian genocide. *Journal of Cultural Geography* 35 (3), pp. 334–361.

Hargreaves, J. 1985. The body, sport and power relations. *The Sociological Review* 33 (1), pp. 139–159.

Harley, J. 1989. Deconstructing the map. *Cartographica: The International Journal for Geographic Information and Geovisualization* 26 (2), pp. 1–20.

Harley, J. 2001. Can there be a cartographic ethics? In P. Laxton (eds.), *The New Nature of Maps: Essays in the History of Cartography*, pp. 197–207. Baltimore, MD: Johns Hopkins University Press.

Harley, J. 2002. *The New Nature of Maps: Essays in the History of Cartography*. Baltimore, MD: Johns Hopkins University Press.

Harper, A. 2010. *The Idea of Islamabad: Unity, Purity and Civility in Pakistan's Capital City*. New Haven, CT: Yale University.

Harrison, T. 2013. *Remembrance Today: Poppies, Grief and Heroism*. London, UK: Reaktion Books.

Harvey, A. 2019. Feminist interventions for better futures of digital games. In B. Beil, G. Freyermuth, and H. Schmidt (eds.), *Playing Utopia: Futures in Digital Games*, pp. 211–234. New York, NY: Columbia University Press.

Harvey, D. 1989. *The Condition of Postmodernity*. London: Blackwell.

Haverluk, T. 1998. Hispanic community types and assimilation in Mex-America. *The Professional Geographer* 50 (4), pp. 465–480.

Hayes, C. 1931. *The Historical Evolution of Modern Nationalism*. New York: Macmillan.

Hayward, K. 2018. The pivotal position of the Irish border in the UK's withdrawal from the European Union. *Space and Polity* 22 (2), pp. 238–254.

Hechter, M. 1977. *Internal Colonialism: The Celtic Fringe in British National Development, 1536–1966*. Berkeley, CA: University of California Press.

Hechter, M. 1992. The dynamics of secession. *Acta Sociologica* 35 (4), pp. 267–283.

Helleiner, E. 2018. *The Making of National Money*. Ithaca, NY: Cornell University Press.

Herb, G. 1999. National identity and territory. In G. Herb and D. Kaplan (eds.), *Nested Identities: Nationalism, Territory, and Scale*, pp. 9–30. Lanham, MD: Rowman & Littlefield.

Herb, G. 2002. *Under the Map of Germany: Nationalism and Propaganda 1918–1945*. London: Routledge.

Herb, G. 2004. Double vision: territorial strategies in the construction of national identities in Germany, 1949–1979. *Annals of the Association of American Geographers* 94 (1), pp. 140–164.

Herb, G. 2018. Power, territory, and national identity. In G. Herb and D. Kaplan (eds.), *Scaling Identities: Nationalism and Territoriality*. Lanham, MD: Rowman & Littlefield.

Herb, G. and Kaplan, D. 1999. *Nested Identities: Nationalism, Territory, and Scale*. Lanham, MD: Rowman & Littlefield.

Herb, G. and Kaplan, D. 2008. *Nations and Nationalism: A Global Historical Overview*. Santa Barbara: ABC-CLIO.

Hernandez, J. 2022. Why Luhansk and Donetsk are key to understanding the latest escalation in Ukraine. *NPR*.

Herod, A. 2010. *Scale*. London: Routledge.

Herzl, T. 1988/1896. *The Jewish State*. New York: Dover Publications.

Hewitt, V. 1994. *Beauty and the Banknote*. London: British Museum Press.

Hewitt, V. 2008. Jammu and Kashmir. In G. Herb and D. Kaplan (eds.), *Nations and Nationalism: A Global Historical Overview*. Santa Barbara: ABC-CLIO.

Hoberman, J. 2020. Sportive nationalism and globalization. In J. Bale and M. Christensen (eds.), *Post-Olympism: Questioning Sport in the Twenty-first Century*, pp. 177–188. London: Routledge.

Hobsbawm, E., and Ranger, T. 2012. *The Invention of Tradition*. Cambridge, UK: Cambridge University Press.

Hoelscher, S. 2003. Making place, making race: performances of whiteness in the Jim Crow South. *Annals of the Association of American Geographers* 93 (3), pp. 657–686.

Hogan, J. 2003. Staging the nation: gendered and ethnicized discourses of national identity in Olympic opening ceremonies. *Journal of Sport and Social Issues* 27 (2), pp. 100–123.

Holford, W. 1962. Brasília: the federal capital of Brazil. *Geographical Journal* 128 (1), pp. 15–17.

Hom, S. 2013. On the origins of making Italy: Massimo D'Azeglio and "Fatta l'Italia, bisogna fare gli Italiani". *Italian Culture* 31 (1), pp. 1–16.

Honey, R. 1999. Nested identities in Nigeria. In G. Herb and D. Kaplan (eds.), *Nested Identities: Nationalism, Territory, and Scale*, pp. 175–197. Lanham, MD: Rowman & Littlefield Publishers.

Hooghe, L., Marks, G. and Schakel, A. 2008. Operationalizing regional authority: a coding scheme for 42 countries, 1950–2006. *Regional and Federal Studies* 18 (2–3), pp. 123–142.

Hoyo, H. 2012. Fresh views on the old past: the postage stamps of the Mexican bicentennial. *Studies in Ethnicity and Nationalism* 12 (1), pp. 19–44.

Hroch, M. 1985. *Social Preconditions of National Revival in Europe: A Comparative Analysis of the Social Composition of Patriotic Groups Among the Smaller European Nations.* Translated by Ben Fowkes. Cambridge, UK: Cambridge University Press.

Huddy, L., Del Ponte, A. and Davies, C. 2021. Nationalism, patriotism, and support for the European Union. *Political Psychology* 42 (6), pp. 995–1017.

Huggan, G. 1989. Decolonizing the map: post-colonialism, post-structuralism and the cartographic connection. *ARIEL: A Review of International English Literature* 20 (4), pp. 115–131.

Huggan, G. 1994. *Territorial Disputes: Maps and Mapping Strategies in Contemporary Canadian and Australian Fiction.* Toronto, ON: University of Toronto Press.

Hunt, D., and Stevenson, S. A. 2017. Decolonizing geographies of power: Indigenous digital counter-mapping practices on Turtle Island. *Settler Colonial Studies* 7 (3), pp. 372–392.

Huntington, S. P. 1993. The clash of civilizations? *Foreign Affairs* 72 (3), pp. 22–49.

Hutchinson, B. and Reevell, P. 2022. What are the Ukraine "separatist" regions at the crux of the Russian invasion? *ABC News*, March 4.

Hutnyk, J. 2005. Hybridity. *Ethnic and Racial Studies* 28 (1), pp. 79–102.

Hwang, J. 2010. The effects of fashion innovativeness and national identity on Korean image clothing preferences. *Journal of the Korea Fashion and Costume Design Association* 12 (3), 17–31.

Hyman, A. 2002. Nationalism in Afghanistan. *International Journal of Middle East Studies* 34 (2), pp. 299–315.

Iamamoto, S. A., Mano, M. K. and Summa, R. 2021. Brazilian far-right neoliberal nationalism: family, anti-communism and the myth of racial democracy. *Globalizations*, pp. 1–17.

Ibhawoh, B. 2008. Nigeria. In G. Herb and D. Kaplan (eds.), *Nations and Nationalism: A Global Historical Overview.* Santa Barbara: ABC-CLIO.

Ibhawoh, B. and Dibua, J. 2003. Deconstructing Ujamaa: the legacy of Julius Nyerere in the quest for social and economic development in Africa. *African Journal of Political Science* 8 (1), pp. 59–83.

Ichijo, A., and Ranta, R. 2016. *Food, National Identity and Nationalism: From Everyday to Global Politics.* Dordrecht: Springer.

Inglehart, R., and Norris, P. 2016. *Trump, Brexit, and the Rise of Populism: Economic Have-nots and Cultural Backlash.* HKS Working Paper No. RWP16-026.

Issacharoff, S., Bursak, A., Rennie, R. and Webley, A. 2019. What is Puerto Rico? *Indian Law Journal* 94 (1), pp. 1–45.

Jackson, J. 1995. Culture, genuine and spurious: the politics of Indianness in the Vaupés, Colombia. *American Ethnologist* 22 (1), pp. 3–27.

Jansson, D., and Koch, N. 2016. Toward a critical geography of sport: space, power, and social justice. In N. Koch (ed.), *Critical Geographies of Sport*, pp. 253–268. London: Routledge.

Jaskot, P. 1996. Anti-Semitic policy in Albert Speer's plans for the rebuilding of Berlin. *The Art Bulletin* 78 (4), pp. 622–632.

Jeffords, S. 1994. *Hard Bodies: Hollywood Masculinity in the Reagan Era.* New Brunswick, NJ: Rutgers University Press.

Jeong, S., and Santos, C. 2004. Cultural politics and contested place identity. *Annals of Tourism Research* 31 (3), pp. 640–656.

Johnson, C. 2018. The changing context of German nationhood. In G. Herb and D. Kaplan (eds.), *Scaling Identities: Nationalism and Territoriality.* Lanham, MD: Rowman & Littlefield.

Jones, R. 2009. Sovereignty and statelessness in the border enclaves of India and Bangladesh. *Political Geography* 28 (6), pp. 373–381.

Jones, R. and Johnson, C. 2016. *Placing the Border in Everyday Life.* London: Routledge.

Jones, R. and Merriman, P. 2009. Hot, banal and everyday nationalism: Bilingual road signs in Wales. *Political Geography* 28 (3), pp. 164–173.

Joseph, M. 1999. *Nomadic Identities: The Performance of Citizenship.* Minneapolis, MN: University of Minnesota Press.

Jüde, J. 2017. Contesting borders? The formation of Iraqi Kurdistan's de facto state. *International Affairs* 93 (4), pp. 847–863.

Judis, J. 2018. *The Nationalist Revival: Trade, Immigration, and the Revolt Against Globalization.* New York, NY: Columbia Global Reports.

Juergensmeyer, M. 2008. *Global Rebellion.* Berkeley, CA: University of California Press.

Kabir, M. 1987. Religion, language and nationalism in Bangladesh. *Journal of Contemporary Asia* 17 (4), pp. 473–487.

Kadman, N. 2008. *Erased from Space and Consciousness.* Jerusalem: November Publishers [In Hebrew].

Kalra, R. 2006. High technology and urban development in Bangalore, India. In J. Gatrell and N. Reid (eds.), *Enterprising Worlds*, pp. 71–81. Dordrecht: Springer.

Kapchan, D. 1995. Common ground: keywords for the study of expressive culture. *The Journal of American Folklore* 108 (430), pp. 479–508.

Kaplan, D. 1989. "Maîtres Chez Nous": the evolution of French-Canadian spatial identity. *American Review of Canadian Studies* 19 (4), pp. 407–428.

Kaplan, D. 1994a. Population and politics in a plural society: the changing geography of Canada's linguistic groups. *Annals of the Association of American Geographers* 84 (1), pp. 46–67.

Kaplan, D. 1994b. Two nations in search of a state: Canada's ambivalent spatial identities. *Annals of the Association of American Geographers* 84 (4), pp. 585–606.

Kaplan, D. 2000. Conflict and compromise among borderland identities in northern Italy. *Tijdschrift voor economische en sociale geografie* 91 (1), pp. 44–60.

Kaplan, D. 2002. Asymmetrical and hybrid identities in the Northern Italian borderlands. In D. Kaplan and J. Häkli (eds.), *Boundaries and Place: European Borderlands in Geographical Context*, pp. 116–140. Lanham, MD: Rowman & Littlefield Publishers.

Kaplan, D. 2017. *Navigating Ethnicity: Segregation, Placemaking, and Difference.* Lanham, MD: Rowman & Littlefield.

Kaplan, D. and Häkli, J. 2002. *Boundaries and Place.* Lanham, MD: Rowman & Littlefield.

Kasianenko, N. 2019. Internal legitimacy and governance in the absence of recognition: the cases of the Donetsk and Luhansk "People's Republics". *Ideology and Politics Journal* 1 (12), pp. 135–149.

Kaufmann, E. 1999. American exceptionalism reconsidered: Anglo-Saxon ethnogenesis in the "Universal" nation, 1776–1850. *Journal of American Studies* 33 (3), pp. 437–457.

Kaufmann, E. 2011. Demographic change and conflict in Northern Ireland: reconciling qualitative and quantitative evidence. *Ethnopolitics* 10 (3–4), pp. 369–389.

Kaufmann, E. 2016. It's NOT the economy, stupid: Brexit as a story of personal values. *British Politics and Policy at LSE.*

Kaya, Z. N. 2020. *Mapping Kurdistan: Territory, Self-determination and Nationalism.* Cambridge, UK: Cambridge University Press.

Keating, M. 2001. *Plurinational Democracy: Stateless Nations in a Post-sovereignty Era.* Oxford, UK: Oxford University Press.

Keating, M. 2012. Reforging the nation: Britain, Scotland and the crisis of unionism. In M. Seymour and A. Gagnon (eds.), *Multinational Federalism.* New York, NY: Palgrave Macmillan.

Kedourie, E. 1993. *Nationalism.* 4th ed. Cambridge, MA: Blackwell.

Kellner, H. 1997. Language and historical representation. In K. Jenkins (ed.), *The Postmodern History Reader*, pp. 127–138. London: Routledge.

Kenny, M., and Luca, D. 2021. The urban-rural polarisation of political disenchantment: an investigation of social and political attitudes in 30 European countries. *Cambridge Journal of Regions, Economy and Society* 14, pp. 565–582.

Kertzer, D. 1988. *Ritual, Politics, and Power.* New Haven, CT: Yale University Press.

Kiljunen, K. 2004. *The European Constitution in The Making.* Centre for European Policy Studies.

Kimmel, M. 2018. The contemporary "crisis" of masculinity in historical perspective. In H. Brod (ed.), *The Making of Masculinities*, pp. 121–153. London: Routledge.

Kitchin, R., and Dodge, M. 2007. Rethinking maps. *Progress in Human Geography* 31 (3), pp. 331–344.

Kitishat, A. R. 2019. William Butler Yeats: the hidden nationalism. *Journal of Language Teaching and Research* 10 (3), pp. 484–488.

Klepek, P. 2021. Internal Sony docs explain how "activities" became a cornerstone for PS5. Retrieved 24 August 2021, from www.vice.com/en/article/5dp34k/internal-sony-docs-explain-how-activities-became-a-cornerstone-for-ps5.

Knight, D. 1977. *A Capital for Canada: Conflict and Compromise in the Nineteenth Century*. Chicago, IL: University of Chicago, Department of Geography.

Koch, N. 2010. The monumental and the miniature: imagining "modernity" in Astana. *Social & Cultural Geography* 11 (8), pp. 769–787.

Kochin, M. 2018. Nations unchained: revolution, empire, and the collapse of the Westphalian order. *Perspectives on Political Science* 47 (1), pp. 38–47.

Kofman, M., Migacheva, K., Nichiporuk, B., Radin, A. and Oberholtzer, J. 2017. *Lessons from Russia's Operations in Crimea and Eastern Ukraine*. Rand Corporation.

Kohn, H. 1944. *The Idea of Nationalism*. New York, NY: MacMillan.

Kosonen, K. 1999. Maps, newspapers and nationalism: the Finnish historical experience. *GeoJournal* 48 (2), 91–100.

Kramer, A. 2022. If war flames in Ukraine, these 2 little enclaves may be the spark. *New York Times*, February 20, www.nytimes.com/2022/02/20/world/europe/ukraine-russia-separatists.html.

Kramer, M. 1993. Arab nationalism: mistaken identity. *Daedalus* 122 (3), pp. 171–206.

Kusek, W. 2015. Transnational identities and immigrant spaces of Polish professionals in London, UK. *Journal of Cultural Geography* 32 (1), pp. 102–114.

Kuzmany, B. 2016. Habsburg Austria: experiments in non-territorial autonomy. *Ethnopolitics* 15 (1), pp. 43–65.

Kwon, S. 2004. Autoexoticizing: Asian American youth and the import car scene. *Journal of Asian American Studies* 7(1), pp. 1–26.

Kymlicka, W. 1996. *Multicultural Citizenship: A Liberal Theory of Minority Rights*. Oxford, UK: Clarendon Press.

Laitin, D. 1998. A theory of political identities. In D. Laitlin and J. Watkins (eds), *Identity in Formation: The Russian-Speaking Populations in the Near Abroad*, pp. 3–35. Ithaca, NY: Cornell University Press.

Laponce, J. 1987. *Languages and their Territories*. Toronto, ON: University of Toronto Press.

Latour, B. 2011. Visualization and cognition: drawing things together. In M. Dodge, R. Kitchin, and C. Perkins (eds.), *The Map Reader: Theories of Mapping Practice and Cartographic Representation*, pp. 65–72. Hoboken, NJ: John Wiley & Sons.

Lauer, J. 2008. Money as mass communication: US paper currency and the iconography of nationalism. *The Communication Review* 11 (2), 109–132.

Lee, J., and Yoon, H. 2017. Narratives of the nation in the Olympic opening ceremonies: comparative analysis of Beijing 2008 and London 2012. *Nations and Nationalism* 23 (4), 952–969.

Légaré, A. 2002. Nunavut: the construction of a regional collective identity in the Canadian Arctic. *Wicazo Sa Review* 17 (2), pp. 65–89. Retrieved June 29, 2021, from www.jstor.org/stable/1409575.

Lehning, P. 1998. *Theories of Secession*. London: Routledge.

Lemon, A., and Williams, O. 1976. *Apartheid: A Geography of Separation*. Farnborough, Hants.: Saxon House.

Leuenberger, C. 2012. Map-making for Palestinian state-making. *The Arab World Geographer* 16 (1), pp. 54–74.

Leuenberger, C. 2019. Crumbling walls and mass migration in the twenty-first century. In R. Jones (ed.), *Open Borders: In Defense of Free Movement*, pp. 177–190. Athens, GA: University of Georgia Press.

Leuenberger, C., and Schnell, I. 2010. The politics of maps: constructing national territories in Israel. *Social Studies of Science* 40 (6), pp. 803–842.

Levine, M. 1991. *The Reconquest of Montreal: Language Policy and Social Change in a Bilingual City*. Philadelphia, PA: Temple University Press.

Lieberman, B. 2013. *Terrible Fate: Ethnic Cleansing in the Making of Modern Europe*. Lanham, MD: Rowman & Littlefield Publishers.

Lijphart, A. 1979. Consociation and federation: conceptual and empirical links. *Canadian Journal of Political Science/Revue canadienne de science politique* 12 (3), pp. 499–516.

Little, J. 2002. Rural geography: rural gender identity and the performance of masculinity and femininity in the countryside. *Progress in Human Geography* 26 (5), 665–670.

Llobera, J. 1990. The historical identity of Catalonia. *Critique of Anthropology* 10 (2–3), pp. 3–10.

Lowe, S. 2014. *Fear and Loathing in La Liga: Barcelona, Real Madrid, and the World's Greatest Sports Rivalry*. New York, NY: Bold Type Books.

Lowenthal D. 1994. European and English landscapes as national symbols. In D. Hooson (ed.), *Geography and National Identity*, pp. 199–232. Cambridge, MA: Blackwell.

Lynch, J. 1973. *The Spanish American Revolutions, 1808–1826*. New York: Norton Press.

Lynch, J. 2001. The colonial roots of Latin American independence. In *Latin America Between Colony and Nation: Selected Essays*, pp. 74–108.

Maake, N. 1996. Inscribing identity on the landscape: national symbols in South Africa. In K. Darian-Smith, L. Gunner, and S. Nuttall (eds.), *Text, Theory, Space: Land, Literature and History in South Africa and Australia*, pp. 145–154. London: Routledge.

Maddison, S. 2020. Separatism as a mode of relations: practicing Indigenous resurgence and nationhood in the 21st century. In S. Maddison and S. Nakata (eds.), *Questioning Indigenous-Settler Relations*, pp. 147–158. Singapore: Springer.

Máiz, R. 2003. Making opportunities: contemporary evolution of Galician nationalism in Spain (1982–2001). *Studies in Ethnicity and Nationalism* 3 (2), 20–34.

Mansvelt Beck, J. 2008. The Basque power-sharing experience: from a destructive to a constructive conflict? *Nations and Nationalism* 14 (1), 61–83.

Marsh, C. 2008. Religion and nationalism. In G. Herb and D. Kaplan (eds.), *Nations and Nationalism: A Global Historical Overview*. Santa Barbara: ABC-CLIO.

Marston, A. 2019. Strata of the state: resource nationalism and vertical territory in Bolivia. *Political Geography* 74, pp. 1–9.

Martin, P. 2002. *Germany: Managing Migration in the 21st Century*. Working Paper CIIP-1.

Mason, W. 2008. European Union. In G. Herb and D. Kaplan (eds.), *Nations and Nationalism: A Global Historical Overview*. Santa Barbara: ABC-CLIO.

Mauter, W. 1998. Churchill and the unification of Europe. *The Historian* 61 (1), pp. 67–84.

Mazzoni, C. 2010. Capital city: Rome 1870–2010. *Annali d'Italianistica* 28, pp. 13–29.

McCallum, R. 1997. Cultural solipsism, national identities and the discourse of multiculturalism in Australian picture books. *Ariel: A Review of International English Literature* 28 (1), pp. 101–116.

McCardell, J. 2008. United States. In G. Herb and D. Kaplan (eds.), *Nations and Nationalism: A Global Historical Overview*. Santa Barbara: ABC-CLIO.

McCloud, D. 1995. *Southeast Asia: Tradition and Modernity in the Contemporary World*. London: Routledge.

McConnell, D. and C. Hurst. 2006. No "Rip van Winkles" here: Amish education since "Wisconsin v. Yoder". *Anthropology & Education Quarterly* 37 (3), pp. 236–254.

McDaniel, A. 1992. Extreme mortality in nineteenth-century Africa: the case of Liberian immigrants. *Demography* 29 (4), pp. 581–594.

McGarry, J. 2012. The United Kingdom's experiment in asymmetric autonomy and the lessons learned. In *Multinational Federalism*, pp. 129–148. London: Palgrave Macmillan.

McKay, I. 2009. *Quest of the Folk: Antimodernism and Cultural Selection in Twentieth-Century Nova Scotia*. Montreal: McGill-Queen's Press.

McLane, J. 2008. India. In G. Herb and D. Kaplan (eds.), *Nations and Nationalism: A Global Historical Overview*, pp. 796–807. Santa Barbara, CA: ABC-CLIO.

McNeill, W. 1998. *Plagues and Peoples*. New York, NY: Anchor.

McRae, K. 1974. Consociationalism and the Canadian political system. In K. McRae, *Consociational Democracy: Political Accommodation in Segmented Societies*, pp. 238–261. Toronto: McClelland and Stewart.

McRoberts, K. 1988. *Quebec: Social Change and Political Crisis*. Toronto: McClelland and Stewart.

Medzini, A. 2017. The role of geographical maps in territorial disputes between Japan and Korea. *European Journal of Geography* 8 (1), 44–60.

Mehari, B. 2018. The right to self-government and representation under Ethiopian Federation: an assessment of the practice and challenges. *Research Journal of Social Science and Management* 8 (6), pp. 88–94.

Meinig, D. 1979. Symbolic landscapes: some idealizations of American communities. In *The Interpretation of Ordinary Landscapes: Geographical Essays*, pp. 164–192. New York, NY: Oxford University Press.

Mihelj, S., and Jiménez-Martínez, C. 2021. Digital nationalism: understanding the role of digital media in the rise of "new" nationalism. *Nations and Nationalism* 27 (2), pp. 331–346.

Mikesell, M., and Murphy, A. 1991. A framework for comparative study of minority-group aspirations. *Annals of the Association of American Geographers* 81 (4), pp. 581–604.

Mill, J. 1998/1859. *On Liberty and Other Essays*. New York, NY: Oxford University Press.

Minghi, J. 1963. Boundary studies in political geography. *Annals of the Association of American Geographers* 53 (3), pp. 407–428.

Mirovalev, M. 2022. Donetsk and Luhansk: what you should know about the "republics". *Aljazeera*.

Miscevic, N. 2008. Philosophy and nationalism. In G. Herb and D. Kaplan (eds.), *Nations and Nationalism: A Global Historical Overview*. Santa Barbara: ABC-CLIO.

Mitra, S. K., and König, L. 2013. Icon-ising national identity: France and India in comparative perspective. *National Identities* 15 (4), pp. 357–377.

Mohr, R. 2006. Territory, landscape and law in three images of the Basque Country. In W. Taylor (ed.), *The Geography of Law: Landscape, Identity and Regulation*, 17–32. New York, NY: Oxford.

Monmonier, M. 1994. The rise of the national atlas. *Cartographica: The International Journal for Geographic Information and Geovisualization* 31 (1), pp. 1–15.

Monmonier, M. 2018. *How to Lie with Maps*. Chicago, IL: University of Chicago Press.

Montpetit, J. 2018. What we've learned so far about the incoming CAQ government: François Legault has displayed both pragmatism and obstinacy. *CBC News*. Last updated: October 6, 2018. www.cbc.ca/news/canada/montreal/what-we-ve-learned-so-far-about-the-incoming-caq-government-1.4850937.

Moore, A. 2016. Ethno-territoriality and ethnic conflict. *Geographical Review* 106 (1), pp. 92–108.

Moore, J. 1984. The political history of Nigeria's new capital. *The Journal of Modern African Studies* 22 (1), pp. 167–175.

Moreno, L. 2007. Federalization in multinational Spain. In M. Burgess and J. Pinder (eds), *Multinational Federations*, pp. 98–119. London: Routledge.

Morisson, V. 2019. Countering allegorical motherhood in Irish and Northern Irish contemporary art: the female body as a tool for resistance. In F. Le Cocq (ed.), *Motherhood in Contemporary International Perspective*, pp. 191–204. London: Routledge.

Mosse, G. L. 1985. *Nationalism and Sexuality: Middle-Class Morality and Sexual Norms in Modern Europe*. Madison, WI: University of Wisconsin Press.

Muasher, M. 2011. *A Decade of Struggling Reform Efforts in Jordan: The Resilience of the Rentier System*. The Carnegie Papers, Washington DC.

Mueller, C. 2006. Integrating Turkish communities: a German dilemma. *Population Research and Policy Review* 25 (5), pp. 419–441.

Muldoon, O. T., Trew, K. and Devine, P. 2020. Flagging difference: identification and emotional responses to national flags. *Journal of Applied Social Psychology* 50 (5), pp. 265–275.

Mulholland, M. L. 2007. Mariachi, myths and mestizaje: popular culture and Mexican national identity. *National Identities* 9 (3), pp. 247–264.

Mulle, E. 2016. Belgium and the Brussels question: the role of non-territorial autonomy. *Ethnopolitics* 15 (1), pp. 105–124.

Munroe, H. 2009. The October crisis revisited: counterterrorism as strategic choice, political result, and organizational practice. *Terrorism and Political Violence* 21 (2), pp. 288–305.

Murphy, A. 1996. The sovereign state system as political-territorial ideal: historical and contemporary considerations. In J. R. Eiser (ed.), *State Sovereignty as Social Construct*, 81–120. Cambridge, UK: Cambridge University Press.

Murphy, A. 1999. Rethinking the concept of European identity. In G. Herb and D. Kaplan (eds.), *Nested Identities: Nationalism, Territory, and Scale*, pp. 53–73. Lanham, MD: Rowman & Littlefield.

Murphy, A. 2002. Brussels: division in unity or unity in division? *Political Geography* 21 (5), pp. 695–700.

Murphy, A. 2013. Territory's continuing allure. *Annals of the Association of American Geographers* 103 (5), pp. 1212–1226.

Murphy-Shigematsu, S. 1993. Multiethnic Japan and the monoethnic myth. *Melus* 18 (4), pp. 63–80.

Myadar, O. 2018. The city, memory, and ideology in Ulaanbaatar. In A. Diener and J. Hagen (eds.), *The City as Power: Urban Space, Place, and National Identity*, 57. Lanham, MD: Rowman & Littlefield.

Mylonas, H. 2019. Nation-building policies in the Balkans: an Ottoman or a manufactured legacy? *Nations and Nationalism* 25 (3), pp. 866–887.

Nagel, J. 1998. Masculinity and nationalism: gender and sexuality in the making of nations. *Ethnic & Racial Studies* 21 (2), 242–269.

Nagel, J. 2008. Gender, sexuality, and nationalism. In G. Herb and D. Kaplan (eds.), *Nations and Nationalism: A Global Historical Overview*, pp. 899–911. Santa Barbara, CA: ABC-CLIO.

Neary, I. 2008. *Burakumin in Contemporary Japan*. London: Routledge.

Nelson, V. 2021. *An Introduction to the Geography of Tourism*. 3rd edition. Lanham, MD: Rowman & Littlefield.

New York Times. 2022. *Maps: Tracking the Russian Invasion of Ukraine*. November 14. www.nytimes.com/interactive/2022/world/europe/ukraine-maps.html.

Ngwa, N., and Akgün, B. 2021. Transformation of national identity, geopolitics and nation building in post-colonial African states: the cases of Cameroon and Senegal. *Journal of Asian and African Studies* 56 (8), pp. 1757–1779.

Nicas, J. and Spigariol, A. 2023a. Lula becomes Brazil's president, with Bolsonaro in Florida. *New York Times*, January 1. www.nytimes.com/2023/01/01/world/americas/bolsonaro-florida-brazil.html.

Nicas, J. and Spigariol, A. 2023b. Bolsonaro supporters lay siege to Brazil's capital. *New York Times*, January 8. www.nytimes.com/2023/01/08/world/americas/brazil-election-protests-bolsonaro.html.

Nietschmann, B. 1994. The fourth world: nations versus states. In G. Demko and W. Wood (eds.), *Reordering the World: Geopolitical Perspectives on the 21st Century*, pp. 225–242. Boulder, CO: Westview.

Nuñez-Mietz, F. 2019. Resisting human rights through securitization: Russia and Hungary against LGBT rights. *Journal of Human Rights* 18 (5), 543–563.

Nurse, K. 1999. Globalization and Trinidad carnival: diaspora, hybridity and identity in global culture. *Cultural Studies* 13 (4), 661–690.

Nwangwu, C., Onuoha, F. C., Nwosu, B. U., and Ezeibe, C. 2020. The political economy of Biafra separatism and post-war Igbo nationalism in Nigeria. *African Affairs* 119 (477), 526–551.

O'Connell, H. 2022. Racism and Confederate monument construction: temporal regimes distinguishing the 1900s, 1960s, and contemporary decades. *Sociology of Race and Ethnicity* 8 (1), pp. 62–78.

O'Sullivan, D. 2006. Geographic information science: critical GIS. *Progress in Human Geography* 30 (6), 783–791.

Office of National Statistics. 2003. Ethnic group statistics: A guide for the collection and classification of ethnicity data. Newport, UK.

Ogden, P. 1989. International migration in the nineteenth and twentieth centuries. In P. Ogden and P. White (eds.), *Migrants in Modern France: Population Mobility in the Later Nineteenth and Twentieth Centuries*, pp. 34–59. London: Unwin Hyman.

Olson, R. 1991. Five stages of Kurdish nationalism: 1880–1980. *Journal Institute of Muslim Minority Affairs* 12 (2), pp. 391–409.

Oluic, S. 2007. *Bosnia and Herzegovina: Identity, Nationalist Landscapes and the Future of the State*. East European Monographs.

Onuoha, G. 2012. Contemporary Igbo nationalism and the crisis of self-determination in Nigeria. *African Studies* 71 (1), pp. 29–51.

Onuoha, G. 2013. Cultural interfaces of self-determination and the rise of the neo-Biafran movement in Nigeria. *Review of African Political Economy* 40 (137), pp. 428–446.

Oriard, M. 2011. Rough, manly sport and the American way: Theodore Roosevelt and college football, 1905. In *Myths and Milestones in the History of Sport*, pp. 80–105. London: Palgrave Macmillan.

O'Rourke, B., and Ramallo, F. 2015. Neofalantes as an active minority: understanding language practices and motivations for change amongst new speakers of Galician. *International Journal of the Sociology of Language* 231, pp. 147–165.

Orridge, A., and Williams, C. 1982. Autonomist nationalism: a theoretical framework for spatial variations in its genesis and development. *Political Geography Quarterly* 1 (1), pp. 19–39.

Osborne, D., Lees-Marshment, J. and van der Linden, C. 2016. National identity and the flag change referendum: examining the latent profiles underlying New Zealanders' flag change support. *New Zealand Sociology* 31 (7), pp. 19–47.

Otiso, K. 2018. Construction of postcolonial Tanzanian national identity. In G. Herb and D. Kaplan (eds.), *Scaling Identities: Nationalism and Territoriality*, 85. Lanham, MD: Rowman & Littlefield.

Ozkirimli, U. 2010. *Theories of Nationalism: A Critical Introduction*. London, UK: Macmillan International Higher Education.

Paasi, A. 1996. *Territories, Boundaries and Consciousness: The Changing Geographies of the Finnish-Russian Border*. London, UK: Wiley-Blackwell.

Paasi, A. 1999a. Nationalizing everyday life: individual and collective identities as practice and discourse. *Geography Research Forum* 19, pp. 4–21.

Paasi, A. 1999b. The changing pedagogies of space: representation of the Other in Finnish school geography textbooks. In A. Buttimer, S. Brunn, and U. Wardenga (eds.), *Text and Image Social Construction of Regional Knowledges*, pp. 226–237. Leipzig, Germany: Institut für Länderkunde.

Paasi, A. 2003. Region and place: regional identity in question. *Progress in Human Geography* 27 (4), pp. 475–485.

Palmer, K. 2016. Homelands as outstations of public policy. *Experiments in Self-Determination: Histories of the Outstation Movement in Australia* 181.

Pan, L. 1994. *Sons of the Yellow Emperor: A History of the Chinese Diaspora*. New York: Kodansha America.

Pantojas-Garcia, E. 2013. The Puerto Rico status question: can the stalemate be broken? *Caribbean Journal of International Relations and Diplomacy* 1 (2).

Parker, G. 1994. *The Times Atlas of World History*. Maplewood, NJ: Hammond.

Parry, K. 2021. The formation of heroes and the myth of national identity. *Sport in Society* 24 (6), pp. 886–903.

Paulin, C. and Woodhouse, K. 2008. South Africa. In G. Herb and D. Kaplan (eds), *Nations and Nationalism: A Global Historical Overview*. Santa Barbara: ABC-CLIO.

Pearce, M. 2008. Framing the days: place and narrative in cartography. *Cartography and Geographic Information Science* 35 (1), pp. 17–32.

Peckham, R. 2000. Mapmania: nationalism and the politics of place in Greece, 1870–1922. *Political Geography* 19 (1), pp. 77–95.

Peluso, N. 1995. Whose woods are these? Counter-mapping forest territories in Kalimantan, Indonesia. *Antipode* 27 (4), pp. 383–406.

Penrose, J. 2002. Nations, states and homelands: territory and territoriality in nationalist thought. *Nations and Nationalism* 8 (3), pp. 277–297.

Penrose, J. 2011. Designing the nation: banknotes, banal nationalism and alternative conceptions of the state. *Political Geography* 30 (8), pp. 429–440.

Perkins, C. 2004. Cartography-cultures of mapping: power in practice. *Progress in Human Geography* 28 (3), pp. 381–391.

Pesic, V. 1996. *Serbian Nationalism and the Origins of the Yugoslav Crisis*. Washington, DC: United States Institute of Peace: 11.

Peters, R. 2006. Blood borders: how a better Middle East would look. *Armed Forces Journal*, June 1.

Peterson, N., and Myers, F. 2016. *Experiments in Self-determination: Histories of the Outstation Movement in Australia*. Canberra, Australia: ANU Press.

Pickles, J. 1992. Texts, maps, and hermeneutics. In T. Barnes and J. Duncan (eds.), *Writing Worlds*. London, UK: Routledge.

Pickles, J. 1995. *Ground Truth: The Social Implications of Geographic Information Systems*. New York: Guilford Press.

Pickles, J. 2011. Texts, hermeneutics, and propaganda maps. In M. Dodge, R. Kitchin, and C. Perkins (eds.), *The Map Reader: Theories of Mapping Practice and Cartographic Representation*, pp. 400–406. Hoboken, NJ: John Wiley & Sons.

Pierce, T. 1996. Philatelic propaganda: stamps in territorial disputes. *IBRU Boundary and Security Bulletin* 4 (2), pp. 62–64.

Pinder, J. 2007. Multinational federations: introduction. In M. Burgess and J. Pinder (eds.), *Multinational Federations*. London, UK: Routledge.

Popov, A., and Deák, D. 2015. Making sense of the "difficult" past: transmission of political heritage and memory-work among young people across Europe. *The Sociological Review* 63, pp. 36–52.

Postero, N. 2007. *Now We Are Citizens: Indigenous Politics in Postmulticultural Bolivia*. Stanford: Stanford University Press.

Presti, L. L. 2021. Mapping coexistence in Italy: reshaping the cartographic feelings of the plural nation. *Political Geography* 85, pp. 102343.

Prins, B., and Saharso, S. 2010. From toleration to repression: the Dutch backlash against multiculturalism. In S. Vertovec and S. Wessendorf (eds.), *The Multiculturalism Backlash*, pp. 82–101. London: Routledge.

Prisco, J. 2020. Ryugyong Hotel: The story of North Korea's "Hotel of Doom". *CNN Style*. www.cnn.com/style/article/ryugyong-hotel-architecture-origins/index.html.

Puerto Rico Report. 2020. How popular is independence in Puerto Rico? July 4, www.puertoricoreport.com/how-popular-is-independence-in-puerto-rico/#.Yc9hiGjMKMq.

Putnam, R. 2007. E pluribus unum: diversity and community in the twenty-first century: the 2006 Johan Skytte Prize Lecture. *Scandinavian Political Studies* 30 (2), pp. 137–174.

Quinn, B. 2003. Symbols, practices and myth-making: cultural perspectives on the Wexford Festival Opera. *Tourism Geographies* 5 (3), pp. 329–349.

Qvortrup, M. 2003. *The Political Philosophy of Jean-Jacques Rousseau: The Impossibility of Reason*. Manchester, UK: Manchester University Press.

Radcliffe, S. 2010. Re-mapping the nation: cartography, geographical knowledge and Ecuadorean multiculturalism. *Journal of Latin American Studies* 42 (2), pp. 293–323.

Raento, P., and Brunn, S. D. 2005. Visualizing Finland: postage stamps as political messengers. *Geografiska Annaler: Series B, Human Geography* 87 (2), pp. 145–164.

Raento, P., Hämäläinen, A., Ikonen, H., and Mikkonen, N. 2004. Striking stories: a political geography of euro coinage. *Political Geography* 23 (8), pp. 929–956.

Ramzy, A. 2021. "They have my sister": as Uyghurs speak out, China targets their families. *New York Times*, November 9. www.nytimes.com/2021/07/27/world/asia/xinjiang-china-crackdown.html.

Randall, J. 1965. *The Career of Philosophy: Volume II: From the German Enlightenment to the Age of Darwin.* New York, NY: Columbia University Press.

Reynolds, A. and Reilly, B. 1999. *Electoral Systems and Conflict in Divided Societies.* Washington, DC: National Academy Press.

Renan, E. 1996/1882. What is a nation? In G. Eley and R. Suny (eds.), *Becoming National.* New York, NY: Oxford University Press.

Resnick, P. 2012. What theorists of nationalism have to learn from multinational states. In M. Seymour and A. Gagnon (eds.), *Multinational Federalism.* New York, NY: Palgrave Macmillan.

Rhodes, M. 2017. Alternative pasts, presents, and futures in Star Trek: historical engagement and representation through popular culture. *Geographical Bulletin* 58 (1).

Ribalow, H. U. 1956. *The History of Israel's Postage Stamps.* Woodbridge, CT: Twayne Publishers.

Ringrose, D. 1998. Capital cities, urbanization, and modernization in early modern Europe. *Journal of Urban History* 24 (2), pp. 155–183.

Rojas, I. 2013. Australian mythscape and national identity: re-imagining Australia. *Creative Approaches to Research*, 6 (2), 22.

Rosecrance, R. 1996. The rise of the virtual state. *Foreign Affairs* 75 (4), pp. 45–61.

Rousseau, J. 1758. *Dissertation on Political Economy.*

Rowe, D. 2004. Antonio Gramsci: sport, hegemony and the national-popular. In R. Giulianotti (ed.), *Sport and Modern Social Theorists*, pp. 97–110. London, UK: Palgrave Macmillan.

Rueda-Saiz, P. 2017. Indigenous autonomy in Colombia: state-building processes and multiculturalism. *Global Constitutionalism* 6 (2), pp. 265–297.

Ruggie, J. 1993. Territoriality and beyond: problematizing modernity in international relations. *International organization* 47 (1), pp. 139–174.

Ryan, S. 1995. Errand into Africa: colonization and nation building in Sarah J. Hale's Liberia. *The New England Quarterly* 68 (4), pp. 558–583.

Rydgren, J. 2003. Meso-level reasons for racism and xenophobia: some converging and diverging effects of radical right populism in France and Sweden. *European Journal of Social Theory* 6 (1), pp. 45–68.

Rydgren, J. and Ruth, P. 2013. Contextual explanations of radical right-wing support in Sweden: socioeconomic marginalization, group threat, and the halo effect. *Ethnic and Racial Studies* 36 (4), pp. 711–728.

Saatci, M. 2002. Nation-states and ethnic boundaries: modern Turkish identity and Turkish Kurdish conflict. *Nations and Nationalism* 8 (4), pp. 549.

Sack, R. 1983. Human territoriality: a theory. *Annals of the Association of American Geographers* 73 (1), pp. 55–74.

Sack, R. 1986. *Human Territoriality: Its Theory and History.* Cambridge, UK: Cambridge University Press.

Safran, W. 2005. The Jewish diaspora in a comparative and theoretical perspective. *Israel Studies* 10 (1), pp. 36–60.

Sancton, A. 1985. *Governing the Island of Montreal.* Berkeley, CA: University of California Press.

Sarantakes, N. 2005. Cold War pop culture and the image of US foreign policy: the perspective of the original Star Trek series. *Journal of Cold War Studies* 7(4), pp. 74–103.

Sasson-Levy, O. 2013. A different kind of whiteness: marking and unmarking of social boundaries in the construction of hegemonic ethnicity. *Sociological Forum* 28 (1), pp. 27–50.

Savage, K. 2009. *Monument Wars: Washington, D.C., The National Mall, and the Transformation of the Memorial Landscape.* Berkeley: University of California Press.

Schatz, E. 2003. When capital cities move: the political geography of nation and state building. *Logos* 94 (4), pp. 109–142.
Schatz, R. and Lavine, H. 2007. Waving the flag: national symbolism, social identity, and political engagement. *Political Psychology* 28 (3), pp. 329–355.
Schechner, R. 2003. *The Future of Ritual: Writings on Culture and Performance*. London: Routledge.
Schendel, J. 2008. Burma. In G. Herb and D. Kaplan (eds.), *Nations and Nationalism: A Global Historical Overview*. Santa Barbara: ABC-CLIO.
Schenkkan, N., and Linzer, I. 2021. *Out of Sight, Not Out of Reach: The Global Scale and Scope of Transnational Repression*. Washington, DC: Freedom House.
Scheppele, K., 2022. How Viktor Orbán wins. *Journal of Democracy* 33 (3), pp. 45–61.
Schetter, C. 2008. Afghanistan. In G. Herb and D. Kaplan (eds.), *Nations and Nationalism: A Global Historical Overview*. Santa Barbara: ABC-CLIO.
Schiller, N., Basch, L. and Blanc-Szanton, C. 1992. Towards a definition of transnationalism. *Annals of the New York Academy of Sciences* 645 (1), pp. ix–xiv.
Schwedler, J. 2012. The political geography of protest in neoliberal Jordan. *Middle East Critique* 21 (3), pp. 259–270.
Scott, C. 2011. *Aboriginal Autonomy and Development in Northern Quebec and Labrador*. Vancouver, BC: University of British Columbia Press.
Scott, D. 2012. How Black nationalism became sui generis. *Fire!: The Multimedia Journal of Black Studies* 1 (2), pp. 6–63.
Scott, J. C. 2008. *Seeing Like a State*. New Haven, CT: Yale University Press.
Searles, E. 2008. Inuit identity in the Canadian Arctic. *Ethnology* 47 (4), pp. 239–255.
Searles, E. 2010. Placing identity: town, land, and authenticity in Nunavut, Canada. *Acta Borealia* 27 (2), pp. 151–166.
Sender, K. 2003. Sex sells: sex, taste, and class in commercial gay and lesbian media. *GLQ: A Journal of Lesbian and Gay Studies* 9(3), pp. 331–365.
Seton-Watson, H. 2019. *Nations and States: An Enquiry into the Origins of Nations and the Politics of Nationalism*. London: Routledge.
Seymour, M. 2012. Introduction: multinational federalism: questions and queries. In M. Seymour and A. Gagnon (eds.), *Multinational Federalism*. New York, NY: Palgrave Macmillan.
Shackle, S. 2017. Trojan horse: the real story behind the fake "Islamic plot" to take over schools. *The Guardian*, February 1, 2017. www.theguardian.com/world/2017/sep/01/trojan-horse-the-real-story-behind-the-fake-islamic-plot-to-take-over-schools.
Shaw, M. 2004. New wars of the city: "urbicide" and "genocide". In S. Graham (ed.), *Cities, War, and Terrorism: Towards Urban Geopolitics*, pp. 141–153. Malden, MA: Blackwell Publishing.
Shelef, N. 2020. How homelands change. *Journal of Conflict Resolution* 64 (2–3), pp. 490–517.
Shlay, A., and Rosen, G. 2010. Making place: The shifting green line and the development of "Greater" metropolitan Jerusalem. *City & Community* 9 (4), pp. 358–389.
Siddiqui, E. 2012. Islamabad: the struggle for Pakistan's identity. *Journal of Undergraduate Research*, pp. 156–172.
Siddle, R. 1996. *Race, Resistance and the Ainu of Japan*. London: Routledge.
Siddle, R. 2008. *The Ainu Indigenous People of Japan*. London: Routledge.
Silvern, S. 2014. Reordering the geography of Indian country. In C. Colten and G. Buckley (eds.), *North American Odyssey: Historical Geographies for the Twenty-first Century*, pp. 69–88. Lanham, MD: Rowman & Littlefield Publishers.
Simpson, A. 2008. *Language and National Identity in Africa*. New York, NY: Oxford University Press.
Sioh, M. 2010. Anxious enactments: postcolonial anxieties and the performance of territorialization. *Environment and Planning D: Society and Space* 28(3), pp. 467–486.

Slater, J. 2019. In Modi's move on Kashmir, a road map for his "new India". *Washington Post*, August 15. www.washingtonpost.com/world/asia_pacific/in-modis-move-on-kashmir-a-road-map-for-his-new-india/2019/08/15/1fff923a-beab-11e9-a8b0-7ed8a0d5dc5d_story.html.

Smith, A. 1979. *Nationalism in the 20th Century*. Canberra, Australia: Australian National University Press.

Smith, A. 1986. *The Ethnic Origins of Nations*. London: Wiley-Blackwell.

Smith, A. 1996. Culture, community and territory: the politics of ethnicity and nationalism. *International Affairs* 72 (3), pp. 445–458.

Smith, A. 2009. *Ethno-symbolism and Nationalism: A Cultural Approach*. London: Routledge.

Smith, W. 1980. Friedrich Ratzel and the origins of Lebensraum. *German Studies Review* 3 (1), pp. 51–68.

Snyder, L. L. 1954. *The Meaning of Nationalism*. New Brunswick, NJ: Rutgers University Press.

Socor, V. 2014. Moscow seeks to upgrade the status of Donetsk-Luhansk in negotiating formats. *Eurasia Daily Monitor* 11 (204).

Sparke, M. 1998. A map that roared and an original atlas: Canada, cartography, and the narration of nation. *Annals of the Association of American Geographers* 88 (3), pp. 463–495.

Sparke, M. 2005. *In the Space of Theory: Postfoundational Geographies of the Nation-state* (Vol. 26). Minneapolis: University of Minnesota Press.

Stamp, G. 2014. Neighbours across the sea: a brief history of Anglo-Irish relations. *BBC News*, April 8 2014, www.bbc.com/news/uk-politics-26883211.

Stewart, A. J., McCarty, N., and Bryson, J. J. 2018. Polarization under rising inequality and economic decline. *Science Advance* 6 (50), pp. 1–29.

Stojanovic, N. 2011. When is a country multinational? Problems with statistical and subjective approaches. *Ratio Juris* 24 (3), pp. 267–283.

Storey, D. 2012. *Territories: The Claiming of Space*. London: Routledge.

Suleiman, Y. 2019. *Arabic Language and National Identity: A Study in Ideology*. Edinburgh, UK: Edinburgh University Press.

Sun, C., and Ha, J. 2020. National identity expressed in Chinese and Korean clothing. *Asian Culture and History* 12 (1), pp. 1–17.

Sutherland, L. 2020. Virtualizing the "good life": reworking narratives of agrarianism and the rural idyll in a computer game. *Agriculture and Human Values* 37(4), pp. 1155–1173.

Tahiri, X. 2018. Politics of enforcing a "multi-ethnic" society in Kosovo: a brief history of constitutional development and state building in Kosovo. *Journal of Danubian Studies and Research* 8 (1), pp. 134–149.

Tamir, Y. 2019. Not so civic: is there a difference between ethnic and civic nationalism? *Annual Review of Political Science* 22, pp. 419–434.

Taras, R. 2008. Nationalism and conflict. In G. Herb and D. Kaplan (eds.), *Nations and Nationalism: A Global Historical Overview*. Santa Barbara: ABC-CLIO

Tazzioli, M. 2015. Which Europe? Migrants' uneven geographies and counter-mapping at the limits of representation. *Movements: Journal for Critical Migration and Border Regime Studies* 1 (2), pp. 1–20.

Tervo, M. 2002. Sports, "race" and the Finnish national identity in Helsingin Sanomat in the early twentieth century. *Nations and Nationalism* 8 (3), pp. 335–356.

Tervo, M. 2004. A cultural community in the making: sport, national imagery and Helsingin Sanomat, 1912–36. *Sport in Society* 7 (2), pp. 153–174.

Thomas, C. 2017. National identity through the connotation of cloth. *The Student Researcher* 4 (1), pp. 1–17.

Thompson, C., and Tian, K. 2008. Reconstructing the south: how commercial myths compete for identity value through the ideological shaping of popular memories and countermemories. *Journal of Consumer Research* 34 (5), pp. 595–613.

Tilly, C. 2002. *Stories, Identities and Political Change*. Lanham, MD: Rowman & Littlefield.

Tivers, J. 2002. Performing heritage: the use of live "actors" in heritage presentations. *Leisure Studies* 21 (3–4), pp. 187–200.

Tomlinson, A., and Whannel, G. 1984. *Five-ring Circus: Money, Power, and Politics at the Olympic Games*. London, UK: Pluto Press.

Toomey, M. 2018. History, nationalism and democracy: myth and narrative in Viktor Orbán's "illiberal Hungary". *New Perspectives* 26 (1), pp. 87–108.

Torpey, J. 2000. *The Invention of the Passport*. Cambridge, UK: Cambridge University Press.

Traub, J. 2016. The party that wants to make Poland great again. *New York Times*, November 2.

Trautsch, J. 2015. The origins and nature of American nationalism. *National Identities* 18 (3), pp. 289–312.

Triandafyllidou, A., 2003. *Immigrants and National Identity in Europe*. London: Routledge.

Troyan, B. 2007. Gregorio Hernández de Alba (1904–1973): the legitimization of indigenous ethnic politics in Colombia. *European Review of Latin American and Caribbean Studies* 82, pp. 89–106.

Troyan, B. 2008. Re-imagining the "Indian" and the state: Indigenismo in Colombia, 1926–1947. *Canadian Journal of Latin American and Caribbean Studies* 33 (65), pp. 81–106.

Tsirbas Y. and Sotiropoulos, D. 2016. Europe at the epicenter of national polities: the attitudes of Greek political elites towards the European Union and the economic crisis. *Historical Social Research* 41 (4), pp. 86–105.

Turnbull, D. 2000. *Masons, Tricksters and Cartographers: Comparative Studies in the Sociology of Scientific and Indigenous Knowledge*. Milton Park, UK: Taylor & Francis.

Tyner, J. 2016. *Landscape, Memory, and Post-violence in Cambodia*. Lanham, MD: Rowman & Littlefield.

Tyner, J. 2020. Urban regeneration and rural neglect: the pall of dark tourism in Cambodia. In N. Wise and T. Jimura (eds.), *Tourism, Cultural Heritage and Urban Regeneration*, pp. 173–185. Berlin: Springer.

Tyner, J., Alvarez, G., and Colucci, A. 2012. Memory and the everyday landscape of violence in post-genocide Cambodia. *Social & Cultural Geography* 13 (8), pp. 853–871.

Unwin, T. 1999. Place, territory, and national identity in Estonia. In G. Herb and D. Kaplan (eds.), *Nested Identities: Nationalism, Territory, and Scale*, pp. 151–174. Lanham, MD: Rowman & Littlefield.

Valková, A. 2014. *A Violence that Did Not Happen: The Peaceful Partition of Czechoslovakia*. Leiden University, Master's Thesis.

Valls, A. 2010. A liberal defense of Black nationalism. *American Political Science Review* 104 (3), pp. 467–481.

Van der Beken, C. 2019. Balancing between empowerment and inclusion: multinational federalism and citizenship rights in Ethiopia. *African Journal of International and Comparative Law* 27 (4), pp. 588–608.

Varshney, A. 2013. How has Indian federalism done? *Studies in Indian Politics* 1 (1), pp. 43–63.

Vertovec, S. 1999. Conceiving and researching transnationalism. *Ethnic and Racial Studies* 22 (2), pp. 447–462.

Vertovec, S. 2004. Cheap calls: the social glue of migrant transnationalism. *Global Networks* 4 (2), pp. 219–224.

Villanueva, N. 2020. *The Athlete as National Symbol: Critical Essays on Sports in the International Arena*. Jefferson, NC: McFarland.

Vittorini, S. 2006. *Representing the Nation: Competing Symbolic Repertoires in India*. (PhD Dissertation). University of London, School of Oriental and African Studies (United Kingdom).

Volcic, Z. 2005. "The machine that creates Slovenians": the role of Slovenian public broadcasting in re-affirming and re-inventing the Slovenian national identity. *National Identities* 7 (3), pp. 287–308.

Vujakovic, P. 2002. Whatever happened to the "new cartography"?: The world map and development mis-education. *Journal of Geography in Higher Education* 26 (3), pp. 369–380.

Wagner-Pacifici, R., and Schwartz, B. 1991. The Vietnam Veterans Memorial: commemorating a difficult past. *American Journal of Sociology* 97 (2), pp. 376–420.

Wainwright, J., and Bryan, J. 2009. Cartography, territory, property: postcolonial reflections on indigenous counter-mapping in Nicaragua and Belize. *Cultural Geographies* 16 (2), pp. 153–178.

Wallach, Y. 2011. Trapped in mirror-images: the rhetoric of maps in Israel/Palestine. *Political Geography* 30 (7), pp. 358–369.

Wang, G. 1991. *China and the Chinese Overseas*. Berkeley, CA: University of California Press.

Warner, M. 2000. *Monuments and Maidens: The Allegory of the Female Form*. Berkeley, CA: University of California Press.

Weber, E. 1976. *Peasants into Frenchmen: The Modernization of Rural France, 1870–1914*. Palo Alto, CA: Stanford University Press.

Webster, D. 2007. From Sabang to Merauke: nationalist secession movements in Indonesia. *Asia Pacific Viewpoint* 48 (1), pp. 85–98.

Weeks, T. 2013. Separatist nationalism in the Romanov and Soviet empires. In J. Breuilly (ed.), *The Oxford Handbook of the History of Nationalism*. Oxford: Oxford University Press.

Weiner, M., and Chapman, D. 2008. *Zainichi Koreans in History and Memory*. London: Routledge.

Westfall, S. 2022. Why are Donetsk and Luhansk in Ukraine's Donbas region a flash point for Putin? *Washington Post*, February 21.

Westwood, S., and Radcliffe, S. 1996. *Remaking the Nation: Place, Identity and Politics in Latin America*. London: Routledge.

White, G. 2000. *Nationalism and Territory: Constructing Group Identity in Southeastern Europe*. Lanham MD: Rowman & Littlefield.

Whyte, C. 2016. Between empire and colony: American imperialism and Pan-African colonialism in Liberia, 1810–2003. *National Identities* 18 (1), pp. 71–88.

Wike, R. 2014. In Europe, sentiment against immigrants, minorities runs high. *Pew Research Center*. Available from: www.pewresearch.org/fact-tank/2014/05/14/in-europe-sentiment-aga inst-immigrants-minorities-runs-high [Accessed 3 January 2023].

Wilkes, J. 2004. Hadrian's Wall. In P. Ganster and D. E. Lorey (eds.), *Borders and Border Politics in a Globalizing World*. Lanham, MD: Rowman & Littlefield Publishers.

Williams, C. 1982. *National Separatism*. Cardiff: University of Wales Press.

Williams, C. and Smith, A. 1983. The national construction of social space. *Progress in Human Geography* 7 (4), pp. 502–518.

Wilson, E. 2005. Gender, nationalism, citizenship, and Nunavut's territorial "house": a case study of the gender parity proposal debate. *Arctic Anthropology* 42 (2), 82–94. Retrieved June 29, 2021, from www.jstor.org/stable/40316648.

Wilson, P. 2021. *Shades of Sovereignty: Money and the Making of the State*. Lanham, MD: Rowman & Littlefield.

Winichakul, T. 1997. *Siam Mapped: A History of the Geo-body of a Nation*. Honolulu, HI: University of Hawaii Press.

Winter, T. 2003. Progressive era. In B. Carroll (ed.), *American Masculinities: A Historical Encyclopedia*, pp. 374–375. Thousand Oaks, CA: SAGE Publications.

Wise, A, and Velayutham, S. 2009. *Everyday Multiculturalism*. Dordrecht: Springer.

Wise, N., and Hall, G. 2017. Transforming Brazil: sporting mega-events, tourism, geography and the need for sustainable regeneration in host cities. In N. Wise and J. Harris (eds.), *Sport, Events, Tourism and Regeneration*, pp. 24–39. London: Routledge.

Wojnowski, Z. 2015. The Soviet people: national and supranational identities in the USSR after 1945. *Nationalities Papers* 43 (1), pp. 1–7.

Wolf-Phillips, L. 1987. Why "Third World"? Origin, definition and usage. *Third World Quarterly* 9 (4), pp. 1311–1327.

Wood, D. 2010. *Rethinking the Power of Maps*. New York, NY: Guilford Press.

Wood, D., and J. Fels. 1992. *The Power of Maps*. New York, NY: Guilford Press.
Wood, M. 2008. Indonesia. In G. Herb and D. Kaplan (eds.), *Nations and Nationalism: A Global Historical Overview*. Santa Barbara: ABC-CLIO.
Xu, X. 2006. Modernizing China in the Olympic spotlight: China's national identity and the 2008 Beijing Olympiad. *The Sociological Review* 54 (2), pp. 90–107.
Yamazaki, T. 2008. Japan. In G. Herb and D. Kaplan (eds.), *Nations and Nationalism: A Global Historical Overview*. Santa Barbara: ABC-CLIO.
Yan, G., and Watanabe, N. 2014. The Liancourt rocks: media dynamics and national identities at the 2012 Summer Olympic Games. *International Journal of Sport Communication* 7(4), pp. 495–515.
Yang, I. C. M., French, J. A., Lee, C., and Watabe, M. 2020. The symbolism of international tourism in national identity. *Annals of Tourism Research* 83, 102966.
Yasir, S., Suhasini, R. and Gettleman, J. 2019. Inside Kashmir, cut off from the world: "living hell" of anger and fear. *New York Times*, August 10.
Zeitz, J. 2015. When America hated Catholics. *Politico Magazine*, September 23.
Zhang, C., Decosta, P., and McKercher, B. 2015. Politics and tourism promotion: Hong Kong's myth making. *Annals of Tourism Research* 54, pp. 156–171.
Zureik, E. 2015. *Israel's Colonial Project in Palestine: Brutal Pursuit*. London: Routledge.

Index

Bold: entries in boxes; *Italics*: figures

FSC
www.fsc.org
MIX
Papier | Fördert
gute Waldnutzung
FSC® C083411

Kolibri 360 GmbH
Ferdinand-Jühlke-Straße 7
99095 Erfurt, Deutschland
produktsicherheit@kolibri360.de